1980

Orang-utan Behavior

Orang-utan Behavior

TERRY L. MAPLE, Ph.D.
School of Psychology
Georgia Institute of Technology
Atlanta, Georgia

Van Nostrand Reinhold
Primate Behavior and Development Series

VNR VAN NOSTRAND REINHOLD COMPANY
NEW YORK CINCINNATI ATLANTA DALLAS SAN FRANCISCO
LONDON TORONTO MELBOURNE

Van Nostrand Reinhold Company Regional Offices:
New York Cincinnati Atlanta Dallas San Francisco

Van Nostrand Reinhold Company International Offices:
London Toronto Melbourne

Library of Congress Catalog Card Number: 79-22233
ISBN: 0-442-25154-8

Manufactured in the United States of America

Published by Van Nostrand Reinhold Company
135 West 50th Street, New York, N.Y. 10020

Published simultaneously in Canada by Van Nostrand Reinhold Ltd.

15 14 13 12 11 10 9 8 7 6 5 4 3 2 1

Library of Congress Cataloging in Publication Data

Maple, Terry.
Orang-utan behavior.

(Van Nostrand Reinhold primate behavior and development series)
Bibliography: p.
Includes indexes.
1. Orang-utan–Behavior. 2. Mammals–Behavior.
I. Title. II. Series.
QL737.P96M36 599′.884 79-22233
ISBN 0-442-25154-8

To the memory of
Richard K. Davenport, Jr.
who really understood orang-utans;
and to Evelyn and Merill Maple
who always understood me.

Van Nostrand Reinhold Primate Behavior and Development Series

AGING IN NONHUMAN PRIMATES, edited by Douglas M. Bowden

CAPTIVITY AND BEHAVIOR: Primates in Breeding Colonies, Laboratories, and Zoos, edited by J. Erwin, T. Maple and G. Mitchell

ORANG-UTAN BEHAVIOR, by Terry L. Maple

GORILLA BEHAVIOR, by Terry L. Maple

CHIMPANZEE BEHAVIOR, by Terry L. Maple

BEHAVIORAL SEX DIFFERENCES IN NONHUMAN PRIMATES, by G. Mitchell

Preface

Orang-utan Behavior is the first of three volumes on the great apes that I began writing in October 1976. I was first contacted by Ashak Rawji of Van Nostrand Reinhold during the annual meeting of the American Association of Zoological Parks and Aquariums, as it had become clear to both of us that there was little formal communication among the many institutions that maintained great apes. The enthusiastic response to my talk on captive orang-utan social behavior was extremely gratifying, and throughout the meeting Mr. Rawji and myself were asked the same general question—Where can I find out more about great apes in captivity? I was unable to give much help on the matter, since the only book on orang-utans (Jantschke, 1972) was available only in German, gorilla books were largely out-of-date (cf. Schaller, 1963), and the vast chimpanzee literature had never really been properly assimilated. Owing to the demands on Mr. Rawji to publish such books, we entered into this ambitious agreement which I may say has scared me immensely. It is my greatest desire that these three books will provide for the reader the answers to many important questions about great ape behavior.

I have attempted to organize the three books so that the general topics covered in each are the same. I have emphasized captive apes although I have not ignored the contributions from the field. Indeed, it is my firmly held belief that one cannot understand captive ape behavior without information from the wild. However, since my greater amount of experience concerns captive animals, and since the problems of captive ape management loom so large, I have elected to direct the greater part of my effort to these issues. For the reader who wishes to know more about wild populations, the illuminating monograph by Rijksen (1978) on orang-utans is especially informative. While the gorilla counterpart has not yet been produced, Schaller's 1963 volume is still a worthwhile source of information on free-ranging gorillas. Wild chimpanzee behaviour is, of course, well documented, in the pioneering volumes of Jane Goodall.

In preparing these books I have been influenced by many works and many people. Among the books which have much impressed me have been Robert and Ada Yerkes' magnificent volume *The Great Apes* (1929). I recommend this fine book (which has recently been reissued by University of Chicago Press) as a good starting point for any student of the great apes. I have also been favorably impressed by the previously-mentioned books of Jantschke (1972) and Rijksen (1978) who have carefully examined orang-utan behaviour in zoos and in the wild respectively. Their recent publications have made my job considerably easier. Conversations and correspondence with Richard K. Davenport, William A. Mason, Jane Goodall, Peter Rodman, Donald Lindburg, Biruté Galdikas, and Robert Sommer have been especially helpful in the formulation of my ideas about apes and captivity. I am especially indebted to my students who have spent so much time with me observing captive apes and discussing the meaning of it all. Evan L. Zucker and Michael P. Hoff deserve special mention for their valuable contributions to my research program, and I am also grateful for the participation of Darlene Clifton, Susan Clarke, Beth Dennon, Charles Juno, Stephanie Puleo, Ronald Schonwetter, Kim Southworth, Sharon Steele, Mark E. Wilson, and Susan Wilson. Without the participation of enthusiastic students, my ambitious great ape research program could not exist. I thank them all. Among my close colleagues I wish to single out Ronald Nadler who has enabled me to collaborate with him without trying to modify my habits and interests. He has been an especially ardent supporter of my students for which I am especially thankful. Both Joe Erwin and Gary Mitchell have been helpful to me by providing friendship, encouragement, collaborative support, and an occasional kick-in-the-pants. I am better for both their candor and loyalty. My friends Mort Silberman and Brent Swenson deserve special recognition for teaching me about the clinical side of apekeeping. They have been models of insightful veterinarians who, like my former colleagues Roy Henrickson and Murray Fowler of the University of California at Davis, are well aware of the genuine value of collaboration among veterinary and behavioral specialists. To Mort I owe a great debt of gratitude for introducing me to a number of valuable research opportunities during the past three years. I am especially indebted to Adelaide Maple for providing support and encouragement, and for typing a considerable portion of the manuscript. In this latter endeavor Celia Sparger also labored on my behalf and I am grateful for her patience, care, and good nature. Brooke Holliman graciously consented to provide numerous drawings of a very high quality and I am pleased to acknowledge her contributions to this book. Jimmy Roberts has always stood ready to assist me, and never ceased to amaze me by his intimate knowledge of the habits of apes. To Frank Kiernan and Larry Fruhwirth I owe numerous beers for photographic assistance. For the use of additional photographic material I thank Jürgen Lethmate, Bobby and Joan Berosini, San Diego Zoo Photolab, and Philadelphia Zoological Society. To the Director of Georgia Tech's School of Psychology, Edward H. Loveland, I owe a special debt of gratitude for his thoughtful support and encouragement.

My research has been financially supported by the following sources: NIH grant RR00165 to the Yerkes Regional Primate Research Center, HD00208 to the Emory University Experimental Psychology Program, an NIH Biomedical Research Support Grant to Emory University, several Faculty Research Grants from Emory University, and more recently from intramural funds from Georgia Institute of Technology. I also wish to acknowledge the assistance of the American Museum of Natural History and the New York Zoological Society which, in the course of supporting an unrelated research project, permitted me access to a quiet place in which to write.

The following individuals commented on earlier portions of the manuscript and I am truly grateful for their criticisms and suggestions, some of which I actually followed: Eric Davis, David Agee Horr, Marvin L. Jones, Lyn Miles, Patricia Scollay, Gene Schmidt, and Brent Swenson. Because of their kind assistance this book has been considerably improved. Needless to say (but I'll say it anyway), they deserve only credit. For any deficiencies which may have been overlooked, I alone must receive the recognition. However, I have honestly tried to be thorough and trust that any omissions or errors will be few and soon corrected. Jane Branch, Leonard Howell, and Kathy Melia cheerfully assisted in the preparation of the index. For permission to use various tables and figures I acknowledge the following sources: Figure. 1-3, Balliere-Tindall Pub.; Fig. 2-1, Tables

2-5, 2-6, 2-7, 2-8, 2-9, 5-5, Academic Press, Inc.; Fig. 4-4, Elsevier-North Holland Press; Fig. 6-3, Tables 5-1, 5-2, 5-3, 6-1, 6-2, S. Karger; Fig. 5-1, Yale Univ. Press; Figs. 5-3, 5-4, 5-5, Primates; Table 6-3, Nelson-Hall Co.

For the duration of this writing effort, I have been permitted free access to the resources of the Yerkes Regional Primate Research Center, the Atlanta Zoological Park, and Kingdom's Three Wild Animal Park. From all of these institutions I have learned a great deal about the nature of the captive experience. I thank Geoffrey Bourne, Fred King, and Steve Dobbs for granting me access. Moreover, it is from the individual inhabitants of these environments, the apes themselves, that I have learned many useful lessons. These books are truly their story.

TERRY L. MAPLE

Contents

Orang-utan Behavior

1 Orang-utan in its Natural Habitat

> There is an increasing tendency toward solitariness as age advances, and the old males spend most of their time alone, whereas the females live with their young. The main reason . . . for isolation or small social groups in the orang-utan is the scarcity of food.
>
> (Yerkes and Yerkes, 1929, p. 136)

Most of the recent research on orang-utan behavior has been conducted in the field. Although both MacKinnon and Rijksen observed captive orangs and mention their observations in their publications, neither investigator set out to *purposefully* study captive orangs. This book is an attempt to integrate the findings of both field and laboratory/zoo researchers. My own research has been conducted on the largest population of captive orang-utans in the world, and while I feel comfortable with the opportunity to generalize, the demands of the captive situtation are clearly different from those of the wild.

Given these hazards, and due to the limitations of my own experience with a captive population, it is essential to begin by outlining some of the important environmental and ecological variables which have affected orang-utan evolution and behavior. With this information at hand, it will be possible to better utilize the details which follow. In each subsequent chapter, if we are to profit from our review of the literature, we must always keep in mind the findings of field workers and the unique adaptations of the arboreal orang-utan.

The problems of generalizing from captive to field setting are legion. The Yerkes recognized the differences in the preparation of their 1929 volume which, by necessity, included more captive than field data.

> We are far from believing that captivity necessarily invalidates naturalistic observation, but it is obviously important to consider in each instance the environmental conditions and circumstances under which psychophysiological data are obtained.

> Captivity may very well transform the attitude, physical condition, and activities of a subject, especially by reason of malnutrition, disease, social deprivation, timidity, dependence on human care, and imitation of human acts . . . Captivity always means opportunity for adaptation, and quite apart from the satisfactoriness of physical conditions, as contrasted with freedom, it has both advantages and disadvantages of scientific inquiry. (p. 112)

As Davenport (1967) pointed out, the Yerkes' landmark volume *The Great Apes* (1929) contained only 91 pages concerning the orang, while the gorilla and chimpanzee were discussed in 145 and 180 pages respectively. Two major survey publications by Devore and Schrier *et al.*, both published in 1965, contained only six to eight pages on the orang while devoting considerable space to the great apes. Clearly, the orang-utan then, as now, was the most mysterious and misunderstood of the great apes. It is my hope that this book will help to bring this interesting animal into a clear and less ambiguous perspective. If we do not yet completely understand its habits, many important questions have now been answered.

To observe orang-utans in captivity, one must be extremely patient. As I will demonstrate, orang-utans emit a variety of complex behaviors, but they are so methodical, cautious, quiet, subtle, and downright slow that many students become discouraged before they even get started. We cannot blame this state of affairs entirely on captivity. As the field records clearly show, orangs are relatively methodical, cautious, quiet, subtle, and downright slow in the wild too! They also sleep a lot, and when they are lying in a motionless heap with their hair in disarray, it seems as if they will remain there in *perpetuity*. Among the potential observers of orang-utans, I was one of the luckier ones. On my first day of observation, within minutes after my arrival, I witnessed a forceful (and lengthy) copulation, and an intense play bout between father and son, events I had not expected to see so soon. The student that I brought with me on that day was flabberghasted since I had so painstakingly convinced him that orangs were little more than red shag-carpeted food processors. How wrong I was to ever harbor such a thought!

I am actually grateful for the species-typical propensities which misled me in the first place. At the right times of the day, orangs move about, play, manipulate objects, fight, and make romance. But, to the delight of the photographer, it is all done *slowly*. In low light there is no need to worry about a slow shutter. And when using motion pictures, there is no need to use slow motion. Orangs are already in slow motion. For the student of orang-utan behavior, these are just a few of the advantages to be gained by selecting this creature for careful study.

In attempting to characterize orang-utan behavior, the greatest single obstacle to progress has been small sample size. Even with the Yerkes population of some 40 orangs, in several instances (cf. Maple *et al.*, 1978; Zucker *et al.*, 1978) we have been forced to publish *case studies* of phenomena previously unreported. While we have tried to be conservative in our interpretation of these events, we will require a great deal more information before we can be certain about the reliability of our findings. Moreover, this same problem exists in the field where observation conditions are much more difficult. For example, our knowledge of orang-utan foraging patterns are based on observations of remarkably few individuals. However, with the recent increase in research, the quantity of information is steadily increasing. For the present, we must rely on our ability to successfully integrate the information which currently exists.

SOME WORDS ABOUT NAMES

Orang-utan is a Malay name which has generally been translated as "man of the forest". A more recent and "with-it" definition has been provided by Biruté Galdikas-Brindamour (1975) who chose to translate the name as "person of the forest." This is an important change of emphasis since, as the reader will see, female orang-utans have not been given a fair share of attention by researchers. In fact, Galdikas herself has remarked to me that she finds adult males "more interesting" than adult-females! But female orangs have not only been ignored, they are also difficult to observe. Despite its importance, maternal behavior has only been successfully studied in captivity.

One other point should be made about *orang-utan* at the outset. Throughout this book, the spelling of the word will differ when other writers are quoted. In his account of *Pongo*, Groves (1971) recommended elimination of the hyphen and use of the common name "orang utan." In all previous publications I have used the hyphenated form, and I will continue the tradition here. As often as not, however, the shorter genus name *Pongo* or orang will be used to designate the subject of this book based on an evaluation of taxonomic usage. If we examine the published literature in the reference section of this book we will find the following distribution of spellings (sans my contributions).

orang-utan : 41
orang utan : 9
orangutan : 8

Thus, the most commonly used spelling is "orang-utan", *with* the hyphen. It is especially interesting to note that *orangutan* (one word) was the spelling selected by Rodman, Galdikas, and MacKinnon in the recently edited volume *The Great Apes* (1979), even though each of them had previously used different spellings. I suggest that we soon make up our minds on this issue.

Also used in the literature is a Dyak name which has been variously spelled *mias, mawas*, and *maias*. I will not use this word except in direct quotations by those who do.

TAXONOMY

The genus *Pongo* has been recognized as stemming from its first use by Lacépède in 1799. This name superseded *Ourangus* as used by Zimmerman (1777) in which he recognized the species *Ourangus outangus*. This name was invalidated in opinion 257 of the International Commission for Zoological Nomenclature (cf. Groves, 1971). Pongo is, respectively, a member of the Order Primates, Suborder Anthropoidea, Superfamily Hominoidia, and Family Pongidae. The pongids include *Gorilla* and *Pan*. *Pongo* includes one species *pygmaeus* composed of the Bornean (*P. p. pygmaeus*) and Sumatran (*P. p. abelii*) subspecies. Historically, the taxonomic labels which applied to orang-utans progressed as follows (cf. Groves, 1971):

1. *Simia pygmaeus* Hoppius, 1763. Type locality erroneously supposed to be Africa. Elliot, 1913, suggested that the specimen probably came from Sumatra rather than Borneo, but Elliot used the name for the Bornean subspecies rather than the Sumatran subspecies, hence by implication restricting the type locality to Borneo, a usage that has prevailed since.

2. *Simia satyrus* Linnaeus, 1766. This name was used for the chimpanzee in 1758.
3. *Orangus outangus* Zimmerman, 1777. Locality unclear.
4. *Pongo borneo* Lacépède, 1799. Locality Borneo.
5. *Simia Agrias* Schreber, 1799. Locality Borneo.
6. *Pongo Wurmbii* Tiedemann, 1808. Locality Borneo.
7. *Pongo abelii* Lesson, 1827. Name wrongly attributed to "Clarke" Abel by Elliot, 1913. Locality Sumatra.
8. *Simia Morio* Owen, 1836. Locality Borneo.
9. *Pithecus bicolor* I. Geoffrey, 1841. Locality Sumatra.
10. *Simia Gigantica* Pearson, 1841. Locality Sumatra.
11. *Pithecus Brookei* Blyth, 1853. Locality Sarawak.
12. *Pithecus owenii* Blyth, 1853. Locality Sarawak.
13. *Pithecus curtus* Blyth, 1855. Locality Sarawak.
14. *Satyrus Knekias* Meyer, 1856. Locality Borneo.
15. *Pithecus Wallichii* Gray, 1870. Allegedly of Blainville, 1818. Locality Borneo.
16. *Pithecus sumatranus* Selenka, 1896. Locality Sumatra.
17. *Pongo pygmaeus* Rothschild, 1904. First use of this combination.
18. *Pithecus wallacei* Elliot, 1913. Allegedly of Blainville, 1839. Not used by Selenka, 1896. Locality Borneo.

Recent opinions rendered by Van Bemmel (1969) and Jones (1969) include the following synonyms for the two recognized subspecies, as listed in Groves (1971): *P. p. pygmaeus*: borneo, agrias, wurmbii, morio, wallichii, brookei, owenii, curtus, knekias, landakkinsis, batangtuensis, dadappensis, genepaiensis, skalauensis, tuakensis, rantaiensis, ladakensis. *P. p. abelii*: bicolor, gigantica, sumatranus, deliensis, abongensis.

The physical characteristics of the genus *Pongo* were described by Groves (1971) as follows:

Ears are small without lobes. Nose is small, without expanded alae. Arms are 200% of trunk length; legs are short, only 116% of trunk length; hands are long, 53% of trunk length; feet are 62% of trunk length (Schultz, 1968). Hallux (first toe) is more reduced than thumb and lacks the terminal phalanx in 75.5% of females and 46.3% of males (Tuttle and Rogers, 1966). Head and body length averages 956 mm. for males and 776 mm. for females, standing height 1366 mm. for males and 1149 mm. for females (Lyon, 1908). Weight averages 75 kg. for males and 37 kilograms for females; a record height is 1800 mm. (Schultz, 1968). Females retain the rounded skull and narrow face of juveniles; most adult males develop a sagittal crest, to which attach massive temporal muscles, and on the face a pair of large laterally projecting cheek-pads (flanges, "blinkers"), which are hard excrescences of connective tissue. Males also develop a long beard and moustache, and the gular sac found in all orangs from shortly after birth becomes enormous and dewlap-like (p. 1-2).

The visual abilities of orang-utans have been studied by Tigges (1963), who found that they exhibit good discrimination of blue, green, yellow, and red, and consistently choose colors over grey. Greater detail concerning the anatomy and physiology of the orang-utan can be found in Groves (1971) and Schultz (1969).

In the wild, identification of different orang-utans is not difficult. Not only do they differ in size, weight, bodily hair, and color, but they may also be distinguished by facial marks, cuts, or scars. (See also Ch. 4) In addition, Rodman (1973) found the shape of the head, height of the hairline, and the direction of hair growth to be reliable indicators of identity. Individual differences in behavior, especially in response to the observer, can also serve to differentiate one subject from another.

DISTRIBUTION

In ancient times, it is likely that the orang-utan ranged throughout Southeast Asia, its remains having been discovered in southern China, Vietnam and Java (Rijksen, 1978; Groves, 1971). Today, the orang is confined to Borneo and Sumatra.

The most recent assessment of the Sumatran orang-utan's distribution was carried out by Rijksen (see Figure 1-1) who surveyed the area by automobile and on foot, and also conducted inquiries on sightings by residents. He concluded that the major concentrations of the animals were as follows:

1. the lowland-and swamp-forests between the Sungei Simpang kiri (southern Alas river) and the Indian Ocean, extending to the north in to the Benkung-and Kluet-areas in the southern part of the Gunung Leuser reserve, and
2. the mountainous forest of the volcanic Kappi-plateau extending to the north into the vast mountain forests of the Serbojadi mountain ridge and the lowland forests of the Jambu Aye tributary. (Rijksen, 1978, p. 34)

In Rijksen's publication, details are given regarding sightings in other regions of Sumatra. However, because of the difficulty in veri-

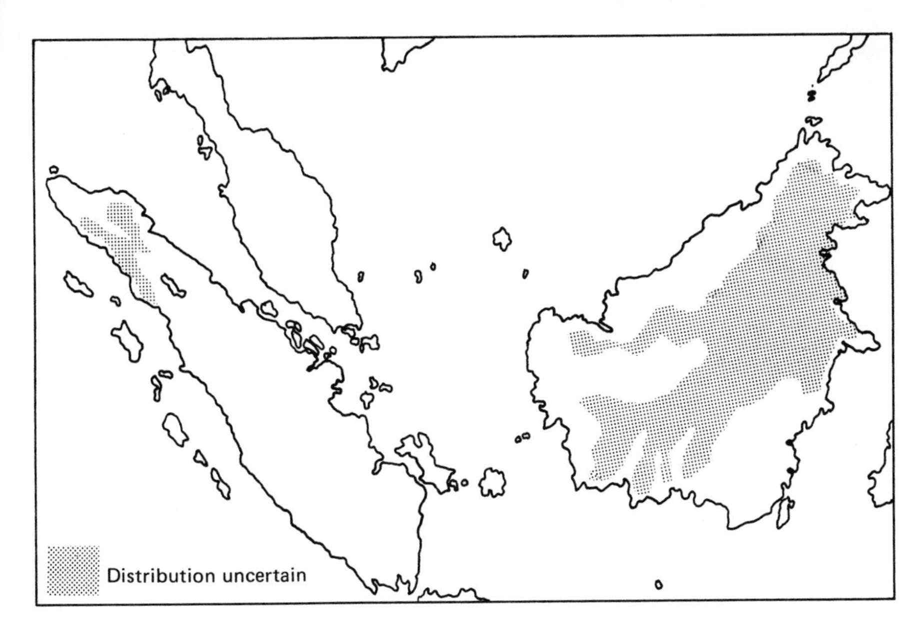

Figure 1-1. The present distribution of orang-utans in Sumatra and Borneo. (After Rijksen, 1978.)

fying these and the few and frequently sketchy reports of other investigations, it is still not possible to determine the distribution of this species. Ominously, Rijksen was forced to conclude that, based on earlier accounts, the distribution range of the Sumatran orang has decreased by 20–30% since the late thirties. He attributes this decrease to massive habitat destruction.

Bornean orang-utans are most abundant in the Ulu Segama Reserve (cf. MacKinnon, 1974) and the Kalimantan study areas investigated by Horr (1972), Rodman (1973), and Galdikas-Brindamour (1975).

Regardless of the subspecies considered, the orang inhabits both swamp/lowland forests (so-called Dipterocarp)* at sea level, and hill/mountain forests at higher altitudes. Surprisingly, orangs do not fear open water, where they have even been encountered up to their waists in water (cf. Hornaday, 1885, as cited in Rijksen, 1978). Anyone familiar with captive orang-utans can report a similar propensity (especially by the young and females of the species) for playing in

* A large family of Asian trees distinguished by characteristic two-winged fruit.

water. In fact, we have observed fully adult males at the Atlanta Zoological Park sitting in their water troughs for prolonged periods as if taking a bath.

EVOLUTIONARY HISTORY

Orang-utans, like gorillas and chimpanzees as we have seen, are members of the family *Pongidae*. The earliest forms of the Pongid family have acquired the label Dryopithecine apes. It appears likely that the ancestor to Pongo was *Dryopithecus sivalensis* whose remains have been found in Northern India and Pakistan (Simons, 1971). It is believed that during the Pleistocene cooling-off period, the orang-utan reached Indonesia by migrating from Indochina across temporary land bridges. Fossil finds indicate that the orang-utans which first migrated to Sumatra (*P. pygmaeus paleosumatrensis*) were about 16% larger than the present day form. The extinct variety which inhabited China may have been as much as 40% larger (Hooijer, 1949).

The structure of its body indicates that the orang has a lengthy history of arboreality. Its large size may have been an adaption for defense against predators. That the prehistoric orangs showed greater sexual dimorphism in canine teeth than the modern forms suggests that orangs may have once been more group-oriented around a single, protective male.

The extinction of the orang on Java, as in other locations, seems to coincide with the presence of *Homo erectus*, who did not inhabit Borneo and Sumatra. That stone age man consumed orang-utans cannot be disputed (cf. MacKinnon, 1974; Medway, 1959).

SPECIES DIFFERENCES

MacKinnon has suggested several ways in which Bornean and Sumatran orang-utans slightly differ, both physically and behaviorally. According to MacKinnon, young Bornean animals usually have bright orange hair. With time, the hair darkens to a chocolate or maroon color, becoming almost black when adult. Sumatran animals tend to be lighter in adulthood and have white or yellow hair on their faces, and in their beard and genital region. The hair of Sumatrans is also said to be "less shiny" but "fleecier" in appearance and to touch. Other microscopic differences have been described in Mackinnon (1973).

The foot of the Sumatran orang is apparently more plantigrade* than that of the Bornean. Moreover, MacKinnon has asserted that the former species tends to more bipedalism. This claim is not supported by any data. MacKinnon, citing van Bemmel, also states that the long, oval face of Sumatran animals differs from the broader "figure eight" shape attributed to Borneans (see Figure 1-2). In addition, there is evidence that the angle between the eyes and the nose is more acute in Sumatrans. The cheek flanges are also said to differ with the Sumatrans exhibiting a "diamond" shape, covered by tufts of short white or yellow "plumose"** hairs. The flanges of Borneans are heavier and less rigid, giving a square appearance, with a tendency to forwardly protrude. Black in color, these "lumpy" flanges or pads contain short and sparse red hairs.

As MacKinnon suggests, Borneans often tend to obesity in captivity.*** However, the assertion that Sumatrans are *less* inclined to obesity cannot readily be substantiated. The Bornean males at Yerkes are not more obese than our Sumatrans, given their identical diet.

Behavioral differences noted by MacKinnon included the use of the hand by Sumatrans when kiss-squeaking, shorter but faster Sumatran long-calls, and longer consortships for Sumatrans. In general, Sumatrans tended to greater sociability and group-cohesion than did the Borneans in MacKinnon's study.

While recognizing the above differences, Rijksen (1978) determined that there existed also two distinct Sumatran body types. He described them as follows:

1. *The dark-haired, long-fingered type*: hair dark brown to maroon; dark brown to blackish skin and facial color; delicate build with slender extremities, long fingers and toes; distinct and well developed thumb and hallux, both capped with a nail; smaller and lighter in weight than the second type.

---

* Walking on the sole, with the heel touching the ground.

** Feathery, plumlike.

*** "Orang-utans teach us that looks are not everything but darned near it. They look awful. Some orang-utans have huge cheek pads and conspicuous laryngeal sacs. Others have worse. The hallux is undeveloped. The female is not so ugly but ugly enough." (Will Cuppy, *How to Tell Your Friends from the Apes,* p. 37).

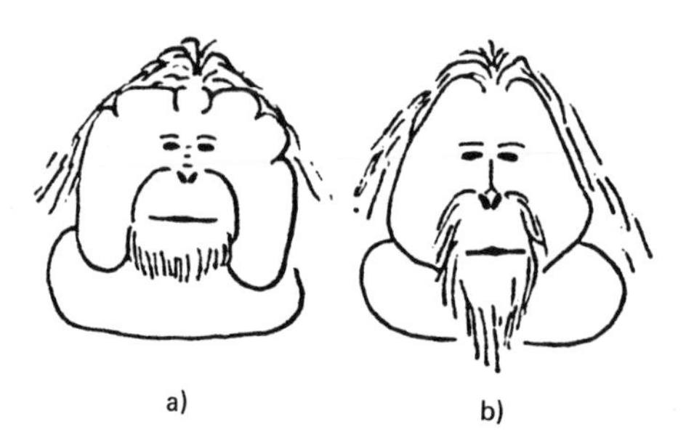

Figure 1-2. Faces of adult male orang-utans showing racial characteristics. (a) Borneo and (b) Sumatra. (After MacKinnon, 1974)

2. *Light haired, short-fingered type*: reddish-cinnamon to rusty-red hair color; light to dark grayish brown skin and facial color; heavily built with stout extremities; short and thick fingers and toes; rudimentary thumb and hallux, sometimes missing the thumb nail, always missing a nail on the hallux.

Rijksen observed numerous intermediates between these extreme forms, but the darker forms exhibited a greater consistency in their characteristics. Like MacKinnon, Rijksen found that behavioral differences correlated with body type. The short-fingered orang-utan was said to be more "friendly, extroverted and, playful." They seemed to be more fearful of novel stimuli and inclined to "bluff" humans and other orangs. The long-fingered type tended to introversion, attacked rather than bluffed, seemed less sociable, and exhibited marked mechanical skills. Long-fingered orangs were also known in Sumatra to be less desirable as pets due to their tendency to aggress.

Given the presence of several extreme and intermediate forms in Sumatra, and the Bornean-Sumatran distinctions, how reliable can bodily differences be in determining the origin of captive specimens? Rijksen's final sentence on these distinctions is well worth quoting here:

It seems therefore rather premature to generalize about the Bornean orang utan in order to give distinctive characteristics distinguishing it from the (variable) Sumatran subspecies. (p.30)

I would add to this, that I am not convinced that it is possible to distinguish subspecific tendencies in captive specimens. There are individual temperament and intellect differences to be sure, but to suggest that captive Sumatrans are more sociable than Borneans is most assuredly premature. Therefore, it would be foolhardy to differentially manage orang-utans on the basis of assumed species differences. The best way to avoid problems is to carefully study the animals which are available, regarding each day and each management change as an experiment. I know of no better justification for a zoo research program than that of taking the guesswork out of management decisions.

STRENGTH

The great strength of the orang has always been appreciated. Most popular accounts of the animal get around to quoting Wallace (1869) on the matter:

> No animal is strong enough to hurt the *Mawas*, and the only creature he ever fights with is the crocodile. When there is no fruit in the jungle, he goes to seek food on the banks of the river, where there are plenty of young shoots that he likes, and fruits that grow close to the water. Then the crocodile sometimes tries to seize him, but the *Mawas* gets upon him, and beats him with his hands and feet, and tears him kills him . . . He always kills the crocodile by main strength, standing upon it, pulling open its jaws, and ripping up its throat . . . The *Mawas* is very strong; there is no animal in the jungle so strong as he, (p.388).

How strong the animal really may be is an open question. The Yerkes wrote the following:

> All authorities who have basis for opinion agree that the orang-utan has remarkable strength and vitality. Especially in the stories of hunters appear frequent accounts of the amazing strength of the animals when hunted, and notably when wounded . . . no less frequent are the accounts of difficulty in killing adults, (p. 135).

The Yerkes' also indicated that the orang's strength, though less than the gorilla, was several times greater than that of a man, and comparable to the chimpanzee. The orang was considered by the Yerkes' to be the most difficult of all primates to kill with a weapon.

If the orang is comparable in strength to the chimpanzee, we can refer to one objective measure as determined by Bauman's (1923; 1926) use of a *dynamometer*. A male chimpanzee weighing 165 pounds and a female of 135 pounds were compared in these studies. The male's strength exceeded the female by about 50%. The two exceeded the strength of a "well-conditioned, well-developed young man" by 4.4 and 3.6 times respectively.

However, Bauman's results must be weighed against a study conducted by Finch (1943) who found that the pulling strength of chimpanzees was only slightly greater than in human beings. As Yerkes (1943) suggested " . . . it seems probable that any advantage which appears is due to continuity and amount of exercise instead of essential differences in physiological process (p. 112)." Objective measures of orang-utan and gorilla strength remain to be conducted, but it seems likely that the strength of both will exceed that of chimpanzees and all but the most highly developed human athletes.

Given their arboreality, I would not be surprised to find that orang males, especially, are even stronger in the upper body than chimpanzees. I particularly remember a struggle for a confiscated hose between two of my students and an animal caretaker on one side, and an adult male orang on the other side. With one hand he held the human trio at bay. It is no wonder that orangs in zoos have been known to turn machine-driven bolts and screws with determined twists of their massive fingers.

DAILY ACTIVITIES

In the wild, orangs rise close to dawn between 0530 and 0800 hours, according to the time zone in which they lived (there is one hour's

Table 1-1. Major field studies of Orang-utan.

Investigator	Species	Location/Duration
Davenport (1967)	Bornean	Sabah/6 months
Horr (1972, 1975)	Bornean	East Sabah (Lokan)/25 months
Rodman (1974)	Bornean	Sabah; East Kalimantan (Kutai Reserve)/14 months
MacKinnon (1974)	Bornean; Sumatran	Sabah (ulu Segama Reserve)/ 16 months; North Sumatra (west Langkat Reserve)/7 months
Rijksen (1978)	Sumatran	North Sumatra (Ketambe in Gunung Leuser Reserve)/3 years
Galdikas-Brindamour (1975)	Bornean	Southeast Kalimantan (Tanjung Puting Reserve)/ over four years

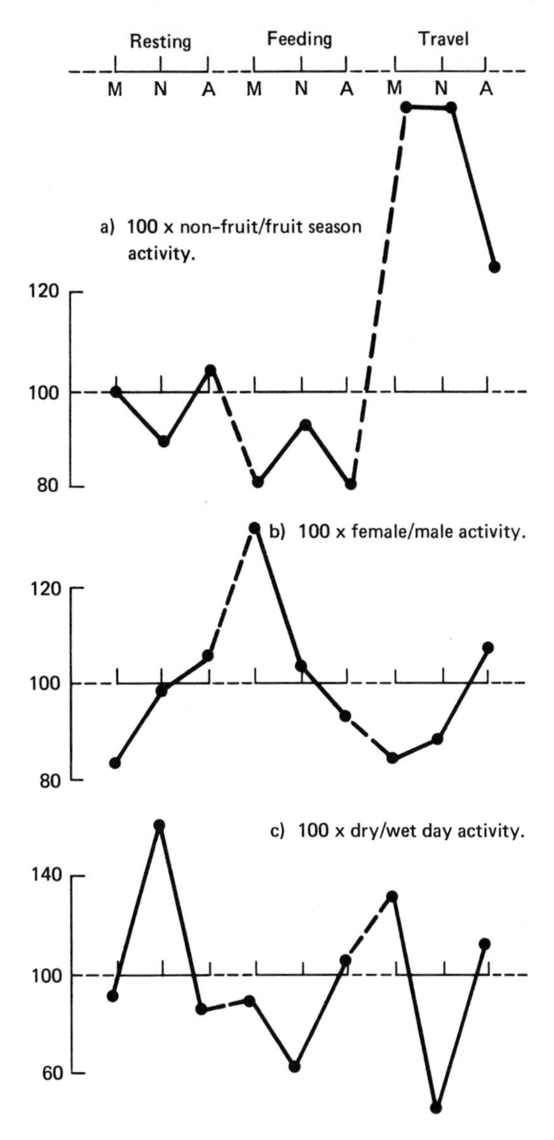

Figure 1-3. Graphs showing effects on orang-utan activity of different factors of season, sex and weather. M = 06.00 hours to 10.00 hours, N = 10.00 hours to 13.00 hours and A = 13.00 hours to 18.00 hours. Ringed point indicates significance $P < 0.05$. (After MacKinnon, 1974.)

difference in Sumatran and North Bornean local time). As Rodman has convincingly shown, adult males spend more time traveling than do other age-sex classes, due to their enormous food requirements.

Both fruiting season and weather exert an effect on patterns of activity. For example, MacKinnon found that orangs spent more time traveling and less time feeding during the absence of fruit. Similarly, more midday resting was observed on *dry* days, and less traveling and feeding. The daily activity pattern is distributed bimodally with peaks at about 0800 and 1500. More feeding takes place in the morning, more travel in the afternoon, while midday is reserved for resting (Rijksen, 1978). Bedtime for orang-utans appears to fall between 1500 and 1800 hours, with little activity at night. However precise these figures may seem, there are many hazards in the plotting of activity patterns as Rijksen has noted:

> . . . great caution is needed in the interpretation of short term, irregularly gathered data. The differences in daily activity patterns of the age and sex classes of orang utans appears to have had insufficient attention. Only a more detailed, long term study on this subject, paralleled by a study of the availability of food resources can provide a firm basis for conclusions on the feeding strategy of this ape. (p. 76)

In captivity, also, orang-utans rise early and, at both Yerkes and the Atlanta Zoological Park, receive a morning feeding. They are quite active at these times, and often engage then in sexual activity. There is a curious correlation between feeding and fooling around! The morning hours between 0800 and 1100 are excellent times to observe social interactions in captive apes. As in the wild, captive

Table 1-2. The daily activity budget concerning feeding and traveling.*

	Feeding	Traveling	Ratio: F/T
Sumatran orang-utan:			
Ketambe area	49.3%	12.2%	4.0
Renun area (from MacKinnon, 1974)	54.1%	20.5%	2.6
Bornean orang-utan:			
Kutai reserve (Rodman, 1973)	45.9%	11.1%	4.1
Segama area (from MacKinnon, 1974)	38.7%	20.9%	1.9
Chimpanzees:			
in 'forest' (Wrangham, 1975)	55.3%	13.4%	4.1
in grass land (Wrangham, 1975)	56.1%	15.8%	3.5

*'Feeding' in the present study includes instances of 'foraging,' i.e. the searching activity employed to feed on small scattered food-items.

'Traveling' in the present study is defined as moving beyond the crown of one particular tree, and thus is distinguished from 'moving,' which encompasses all changes of locations by means of locomotor activity. (After Rijksen, 1978.)

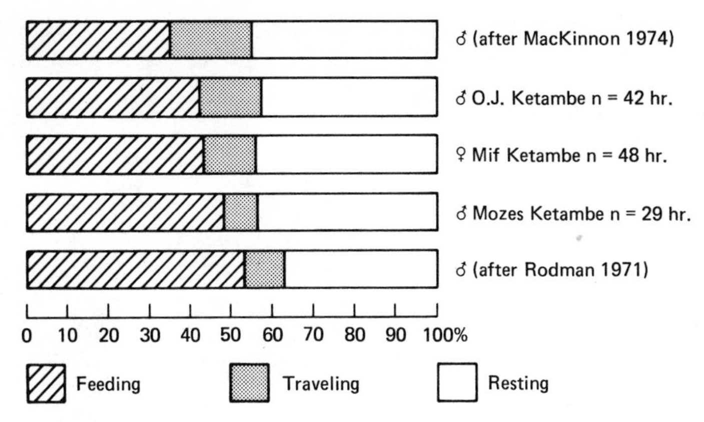

Figure 1-4. Comparison of activity profiles of orang-utans (Bornean data after Rodman, 1971 and MacKinnon, 1974). (Figure after Rijksen, 1978.)

orangs also nap at midday, becoming active again at about 1400–1500 hours. In our studies, we always sample during these activity periods, knowing that other factors which affect interaction will still be reflected in the magnitude of the response in question.

FEMALE MOVEMENT

An area of apparent controversy has been the size and stability of the home ranges of adult females. Both Rodman and Horr found that adult females were resident throughout the year, residing in small home ranges approximately 0.65 km.[2] Contrary to these findings, MacKinnon observed a considerable degree of mobility in the animals within his study site. A consistent finding among the three observers was the overlap in the home ranges of females. The balance of opinion in the three studies also favors the view that there is overlap in the home ranges of adult males, which maintain ranges which are twice those of females, although some boundaries were apparently defended. A common conclusion was that Bornean orang-utans were widely dispersed in small subgroups in order to successfully exploit the dispersed food sources which they preferred. The differences in reported dispersal patterns were thought by MacKinnon to be due to methodology differences in the three studies. Where there have been differences in the size of study populations, these

Figure 1-5. Alfred Russel Wallace made early observations on orang-utans.

differences may have also affected dispersal. Obviously, the methods used and the characteristics of the study regions account for the behavior of individuals and thus, on this basis alone, comparisons are hazardous and not likely to yield complete agreement. The availability of food will most certainly determine the size of the home range, as will the number of competitors, and other pressures exerted on the forest food resources.

TERRITORY MAINTENANCE IN MALES*

How long orang-utans are able to keep territories, or how long a given individual may inhabit a specific home-range is currently unknown. Presumably, factors such as population density, human encroachment, food supply, and individual health will determine the stability of territory and home range. An interesting statement by Carpenter (1937) gives us one example of such stability:

> . . . a large male of the group observed near Tapa Toean was said to have ranged in a given territory for at least the last twenty years. Whether or not these great apes migrate and for what distances is an important consideration in relation to the problem of wild life protection and the location of proposed reservations and their size (p. 16).

Rodman (1979) from an intensive study of the movements of one mature male, has convincingly argued that male distribution and ranging patterns is an adaptation to the spatial and temporal distribution of ovulating females. Since such females are scarce, mature males may engage in the following alternative strategies:

1. constant association with one adult female;
2. expansive travel to contact many ovulating females;
3. territorial defense.

According to Rodman, the latter strategy involves the least amount of energy, for the greater reproductive advantage. The extension of this argument suggests that stronger male competition has resulted in marked sexual dimorphism.

Although the estimates of home range vary from Rodman's 0.42 km^2 low to Galdikas' (1979) high of 5–6 km^2, daily ranging patterns

*Horr (1977) has noted that male home ranges may be too large to be effectively defended.

were estimated from 305 m/day (Rodman) to 800 m/day (Galdikas). It would be interesting to see whether captive orang-utans ever travel so much in their enclosures, even though it would have to be done in circles.

It should be noted, however, that as MacKinnon (1979) has asserted, orangs are not "trapped" within their ranges. The most important constraint upon spatial behavior is apparently food not fear. MacKinnon considers orang society as extremely flexible, whereby spatial distribution is directly influenced by availability of food. Furthermore, he has argued, orangs must meet and therefore recognize all others that share its range, thereby determining its primary social contacts.

Apparently, fierce territorial fights among adult males do occur in the wild. Galdikas (1979) noted that mature males can be distinguished by their disfigurements, presumably the result of conflicts:

> Digits seemed especially vulnerable; some males had parts of fingers missing or stiff fingers that were never observed bending . . . misshappen or broken fingers and toes. The right eye of one male was missing. Another male had a chunk of flesh missing from his upper lip . . . (p. 216)

The intensity of fighting, according to Galdikas, is determined by the presence or absence of females. In the latter instance, fighting is considerably less likely.

Of course, overt fighting may be considerably reduced by the ritualized long-call, which indicates the male's location. This vocalization is made up of a series of grumblings and bellowings (cf. Galdikas; 1979; MacKinnon; 1974) which can last four minutes or more. Long calls occur daily, generally two to five times (Galdikas, 1979), and are frequently accompanied by the audible dropping or tearing of branches. Although long distant calling elicited no response in others, males in consortship have been reported to be unusually responsive (Galdikas, 1979).

Field workers have also argued that the long call functions to attract females while simultaneously repelling males. With both functions; the male identifies his location by calling.

MacKinnon (1974, 1979) has observed that in both Borneo and Sumatra, orangs were organized into a "loose type of society". He used the term *community* to describe Sumatran orangs which ranged over the same forest areas, exhibited coordination in travel patterns, and were frequently distributed around single mature males. In Borneo, MacKinnon found the term to be somewhat less appropriate. The slight difference in Sumatran and Bornean *community* has been frequently generalized by zoo workers into full-fledged species differences in sociability. The evidence from both captive and field studies does not yet support this notion.

FOOD HABITS IN THE WILD

Orang-utans are primarily fruit-eaters in the wild, but they have been known to eat a variety of other vegetation including leaves, shoots, flowers, epiphytes, lianas, woodpith, and bark (MacKinnon, 1974). In addition, orangs sometimes eat soil, insects, and birds' eggs. Lists of the foods which were eaten by orangs can be found in the published studies of MacKinnon and Rijksen. As MacKinnon noted, orangs apparently prefer uncommon foods which are considerably dispersed. As a result, orangs must always be on the move in search of their preferred foods. Outside the fruit seasons, orangs are especially active. In the course of their travels it was apparent to both MacKinnon and Rijksen that orangs moved along natural boundaries, such as ridges and streams. These traditional arboreal routes were called "highways" by MacKinnon. The facility with which orangs located new food sources is most remarkable.

> In view of the unpredictable nature of fruit availability orang-utans seemed to have an uncanny ability of arriving at the right place at the right time. I rarely discovered a new foodsource before they did . . . Some fruits, e.g. durians and figs, have a strong smell which may attract animals. The noisy commotion of feeding doves, hornbills, gibbons and monkeys may also guide orang-utans to fruit trees at times. Whatever their methods, they are certainly very efficient at finding small, rare, dispersed sources of food . . . In view of this, the relatively solitary way of life of the orang-utan seems to be well adapted, (MacKinnon, 1974, p. 31).

Because of their arboreality and, all too frequently, their aversion to humans, wild orang-utans are not generally easy to study. For some activities, however, there are distinct advantages in studying orang-utans as compared to other arboreal primates. As Rodman (1977) has argued, since orangs are large, conspicuously colored, and deliberate in their movements, their feeding behavior is relatively easy to study. Moreover, because they do not congregate in large groups, it is possible to focus on one or several animals with few distractions.

Rodman also determined that, based on a 12 month average, orangs fed 53.8% of the time on fruit, 29.0% on leaves, 14.2% on bark, 2.2% on flowers and 0.8% on insects. At least 50 different species of trees and vines were selected by the orangs in East Kaliamantan, perhaps as many as 80. Rodman noted that MacKinnon's orangs in Sabah fed on 105 distinct species, suggesting that this difference might account for other differences in the animals found in the respective study sites.

The solitary life-style of orang-utans has been explained by Rodman (1977) as being due to their dependence on dispersed patches of fruits without threat of predation. Such a condition favors solitary over aggregating animals. Moreover, male behaviors are further influenced by competition for females, while the two sexes are segregated by differential exploitation of the resources. From an admittedly small sample, Rodman cautiously concluded that adult males altruistically selected "less" preferred food items in order to avoid competing with females and young that he may have sired.

Perhaps the most detailed study of feeding to date is that of Rodman. In this paper can be found an elegant discussion of the better methods to be employed in studies of feeding. In recording general behavior patterns, Rodman used the following categories: Feeding, Traveling, Resting, Nest-building, and Display. Among the latter were included the small units of vocalizing, tree shaking, branch-breaking, and vine rattling. These behaviors were recorded while following orangs, but Rodman's methods permitted the following:

> . . . observations of feeding behavior consist of the times of initiation of bouts of feeding, their durations in whole minutes, and a set of descriptions of feeding within bouts including identification of food type and food source (if known), rates of feeding, height above ground, postures utilized, and means of support while feeding. (p. 386)

In feeding near the ends of tree crowns, the main location for orang-utan feeding, Rodman described two primary feeding postures, an upright orientation, and a horizontal orientation. In the vast majority of instances, orangs attached themselves to trees by several appendages in order to provide adequate support. Obviously, due to their great weight, adequate support postures are adaptations of some value.

Rodman's recently published account of activity was broken down as follows:

Feeding	45.9%
Resting	39.2%
Traveling	11.1%
Displays	2.7%
Nest-building	1.0%

Orang-utans consume food with the great facility provided by their mobile lips and cavernous mouths. As MacKinnon has shown, small food items such as leaf shoots and small fruits are plucked by the lips and ingested immediately. Because orangs bend and often break branches during feeding, damaged trees are often a sign to the observer that the red apes have been there. With binoculars, MacKinnon was able to spot regular orang feeding areas from far away. An excellent description of the use of lips in feeding is provided in the following excerpt:

> Animals can shell and clean several fruits in their mouths at the same time. Periodically they extend the lower lip with the fruit arranged on it and squint down at them. They then either spit them out or take them back for more rasping (MacKinnon, 1974, p. 44).

In the wild, social feeding takes a somewhat different form than in captivity where food is less abundant. In the zoo, priority of access means that one animal gets its food sooner than others. In fact, a distinct feeding order develops in captivity, rarely deviating from day to day. In general, orangs tend to feed separately and quietly without much fighting. At Yerkes and at the Atlanta Zoo, purposeful regurgitation of food is a common, and repetitive occurrence. (cf. chapter 7). Favored foods are generally eaten first.

Many primate studies have discussed the relationship between size and access to feeding sites (cf. Grand, 1972; Kummer, 1971). Rodman (1973) has proposed that orang-utan social organization is affected by differences in the feeding requirements of larger males and smaller females. Thus, greater ranges and, hence, spatial separation, may be required for the adult male orang-utan. For example, Rijksen (1978) has argued as follows:

A large animal is able to harvest a relatively larger proportion of a non-durable fruit crop than a smaller one. Thus, especially when excess energy can be stored in the form of fat, being of large size can be seen as a particular strategy under particular environmental conditions. That is, temporary availability of large quantities of food, which occur regularly in time, and a special position with regard to predatory pressure. (p. 125)

Regarding the *relative selectivity* of the orang, Rijksen (1978) has suggested that the long tailed macaque and the orang-utan are more opportunistic than are the gibbon and the leaf monkey respectively, which are considered to be more *selective* in their food habits. As Rijksen has also pointed out, it is "not selectivity as regards different food-plant species, but selectivity in quality of fruits within the same crop of a fruit tree species, that seems to play the major role." (p. 176)

It is interesting to note that, as Rijksen (1978) points out, of the three main primate groups in the Ketambe area of Sumatra (leaf monkeys, macaques, apes), only *one* species has adapted to primarily consume the ever-abundant rain forest leaf materials, whereas at least five primates and numerous other animals compete for other less abundant food items. As Rijksen has concluded:

The tropical rainforest ecosystem consists of an immensely complicated network of subtle relationships between the many different components . . . The synecology of the primate species of the Ketambe area certainly deserves a much more detailed approach . . . it requires a much more detailed analysis of the respective species' feeding behavior, their energy balance, as well as a close monitoring of natural changes in the habitat conditions from day to day for each species. Yet, notwithstanding the paucity of firm conclusions . . . the orang-utan is exposed to a considerable degree of food competition from a variety of mammals and birds. This particularly concerns his staple food, figs. (p. 133)

FEEDING IN CAPITIVITY

At the Yerkes Primate Center, adult orang-utans have been traditionally fed a diet consisting of *chimcrackers*, steamed white and sweet potatoes, raw cabbage, cooked beans with molasses, steamed rutabaga, oranges, liquid milk with veterinary yeast*, rolled oats,

* Imported from England, Vetzyme (Phillips Yeast Co., Park Royal, London.)

Figure 1-6. Early drawing of adult male orang-utan as it appeared in Mivart's *Man and Apes* (1873). Note walking stick.

and *grits*! Chimcrackers* contain graham flour, cracked wheat, corn meal, oatmeal, powdered whole milk, peanut butter, ground raisins, dark molasses, burnt ash and table salt (Bourne, 1972). Dr Bourne and others have argued that successful reproduction depends on proper nutrition. However, as this book goes to press, the Yerkes Primate Center has initiated a total biscuit diet for some great apes. It appears that such a diet does not affect the general physical health of the apes, at least in the short run. However, the lack of edible vegetation must be considered as one more psychological insult to animals which already encounter few objects in their daily lives. As Lang (1972) points out, a wide variety of food items provide scope for imagination and occupation.

For pregnant females, the elaborate diet which I have outlined as above has been supplemented by wheatgerm and corn oil supplement. Some zoos have given Ovaltine in milk during pregnancy. Information on the diet of nursery-reared infants is presented in Chapter 5.

A hierarchical order of feeding among captive orang-utans has also been noted by Brandes:

> If a male, let us say the head of the family feeds, the female sits down near him and counts, so to speak, every bit into his mouth. This behaviour is so distinct that spectators often expressed their resentment at the selfishness of the male! But I came to the conclusion that the female was neither jealous nor greedy for his food. If I approached her in such situations with special tid-bits, grapes for instance—even putting these right between her lips—she refused to accept them. It was her 'duty' to attend to the male first. Only after he completed his meal, did she think of her own food. (Brandes, cited in Harrisson, p. 153)

Harrisson noted about this statement that there were numerous human beings in the world who have practiced similar eating habits. As she asserted (p. 153): "It seems a convenient and respectable way both to care for and control the male, and gives him a feeling of superiority at the same time." The key to this argument is the word *control*. The orang does not seem inclined to *actively* share food, even with infants. The first solid foods an infant eats are those tidbits it can take from the mother's hand or mouth. The infant gets increasingly good at this as its perceptual and motor skills mature, but the

* 3.95 cal/g.; 13.3% from protein, 20.9% from fat, 65.8% from carbohydrates.

mother does not generally encourage its efforts. However, I have observed several instances of what I would call *instrumental food-sharing* where a female *actively* holds out food to a pursuing male. It has never been clear to me whether the male intended to take her food or copulate with her on these occasions, but upon receiving the handout, he always ceased his pursuit. In my opinion, the female gave up the food to deter the male's advance.

However, despite my guarded pessimism, there are many reports of active food-sharing on the part of mothers. In some instances, this may be due to overzealous interpretation, but it doubtless occurs at least in some mothers. Passive food sharing, whereby one animal is permitted access to another's food, has been observed among some of our young subjects, and also between adult heterosexual pairs.

NESTING

Orang-utans construct sleeping nests daily for napping, and for each evening's extended sleeping session. In fact, the occurrence of nests has been used by field workers to calculate the number of animals which inhabit a given area. As Rijksen has shown, nests can remain visible from two weeks to more than a year. However, from Rijksen's 1978 data, it appears that a modal number of nests are visible for from 2½ to 3½ months. The standard method for calculating orang numbers from nest numbers has been to use the following equation:

$$d = N/xc$$

where d (density) = no. of animals/Km^2

- N = number of nests located
- x = distance surveyed
- c = product of the observation distance on either side of the tracks × the number of nests built by an individual orang during a nest's average life.

The variability in past estimates may well be due to the different estimates about nest life, which have ranged from two (Kurt, 1970) to six months (Schaller, 1961).

Orangs build at least one nest per day. The relative permanence of these nests varies according to at least the following variables as suggested by Rijksen: construction technique, the animal's size and weight, the builder's "mood", location and characteristics of the tree, and weather. A final fortuitous factor is whether other animals happen to prematurely destroy the nest in search of insects.

Age is apparently a factor in nest-building as Rijksen noted that young animals tend to build a large number of nesting or "play" nests per day. Existing nests can also be re-used and, in some cases, rebuilt by other orangs. While other workers have been satisfied with the assumption that orangs build one nest per day, Rijksen calculated the average number as 1.8 per day with a recorded range of zero to six. In an elegant discussion of the problems and potential errors inherent in the nest estimation process, Rijksen concluded that:

> In practice, nest counting surveys can only give an answer to the question whether or not orang utans are present in a certain area. However, in conservation practice, nest counting surveys may have some value in comparing densities . . . e.g. before and after disturbance. Such surveys provide a comparatively easy method to assess the effects of disturbance in particular when time is a limiting factor. (p. 42)

According to Rijksen, a nest is generally constructed so that the animal will have an unobstructed view of a large segment of the forest. This generalization apparently holds true for both Bornean and Sumatran orangs. MacKinnon (1974) has argued that "good feeding sites, salt licks (Sumatra) and geographical features such as ridge junctions observed that concentrations of nests which often occurred on western-facing slopes may have been due to the warmth of the sun, shelter from evening winds, and the extensive view (p. 48). Another interesting factor of influence may be the presence of other nests. It seems to me that the psychological principle of *social facilitation* could account for this propensity in certain instances. However, MacKinnon seems to favor the notion that orangs actually seek out known nesting sites, used by themselves or others, perhaps because of the correlation between nests and the favorable advantages which they bestow. If a preferred fruit tree is particularly attractive, Sumatran orangs may reuse their nest on consecutive days, but they generally demonstrate their preference for particular locations by returning to nests at two–to eight-month intervals. As we have seen, captive animals, with limited space, tend to perseverate in their preferred sleeping locations. This captive propensity is reflected in Rijksen's report of a rehabilitant orang which reused a nest for 18 consecutive nights. While MacKinnon (1974) concluded that orang-utans frequently made evening nests in or near the last located food tree, Rijksen argued otherwise:

> We namely had the impression that one possible reason for the reluctance of most orang utans to build their sleeping nests in attractive places such as preferred fruit trees, could be the risk of agonistic encounters if they did so. The risk for agonistic encounters with conspecifics of higher social status seems evident, but there may also be a risk of agonistic encounters with other species, in particular with man, who regards several of the 'preferred' fruit tree species of the orang utan as property and may kill competitors. (p. 151)

The discrepancy between the two generalizations may be due to differences in the likelihood these suggested limiting factors accord to the respective study sites, or to unavoidable sampling bias by one, the other, or both of the observers.

In the wild, orangs generally bed down about 30 minutes before sunset, retiring earlier during bad weather (MacKinnon, 1974). In instances of heavy rain, the day nest refuge often became the night nest as well. Orang-utans apparently rarely leave their nests after dark, but since human observers have not been able to effectively study orangs at night, their relative propensity to do so is currently unknown. The amount of available moonlight may be a factor which determines night-time activity, as suggested by the observations of Harrisson (1969).

The time at which orangs leave their nests in the morning varies considerably, but the average recorded by MacKinnon was 0630. Again, bad weather caused animals to remain late in the nest.

Bernstein (1969) studied the nest building habits of captive orangs at the Yerkes Regional Primate Research Center. From this research, it is clear that experience is a factor in the development of nesting behavior.

He observed two young orangs between 2½ and 3½ years of age, and 13 adult orang-utans, each of which was given nesting material either for the first time in their lives, or for the first time since their early infancy in the wild. In all cases, the orangs became increasingly adept at nest building as they gained experience with the material. The motor patterns associated with orang nest building were essentially the same as those of wild orangs, and quite similar to the nest-building behaviors of captive chimpanzees and gorillas. Despite what appears to be a species-typical propensity to build a nest, Bernstein concluded that experience was important, and that nest building had to be learned by apes early in their lives, as the following statement reveals:

> The phenomenon of nest construction, as described for all three Great Apes, is still imperfectly understood despite the growing body of field and laboratory studies. It is possible that a principle such as imprinting may be required to explain apparent critical ages in learning to produce nests, although the presence of the requisite motor skills in adult laboratory born animals may mean that a shaping procedure could also be used to teach these animals to build nests. Spontaneous nest construction, however, may depend on early exposure to nest construction by the mother. (p. 402)

Lethmate (1977) also examined nest-building in captivity and found that an isolation-reared male orang emitted "typical nest-building" behaviors at the age of fourteen months. However, this early propensity for nest building was carried out with inappropriate materials such as blankets and food. Over a twelve month period, this subject also treated twigs, wooden blocks, and tools as nest-building material.

Although we will examine orang manipulativeness in greater detail in Chapter 6, their technique in building secure sleeping nests can be considered here. In view of Bernstein's findings on the acquisition of nest building in captive orangs, it is useful to compare the observations of both MacKinnon and Rijksen on wild orangs.

In MacKinnon's 1974 study the animals were observed to move in a circle bending the branches inward with hands and feet. The result was a "concave springy platform" which took two to three minutes to construct. MacKinnon determined further that branches were used in the following ways (p. 49):

1. *Rimming*, whereby branches were bent horizontally to form the nest rim and held in place by other bent branches.
2. *Hanging*, in which a branch is bent down into the nest to form part of the nest bowl.
3. *Pillaring*, in which branches are bent over from beneath the nest to hold rim branches in place and give extra support.
4. *Loose*, whereby a branch is snapped from the tree and put into the nest bottom or put into place above as part of the roof.

Captive and rehabilitant animals often make nests on the ground but, as we have seen, the essential mode of nest building does not seem to differ among orang-utans regardless of habitat.

Overhead shelters were often constructed by the orang-utans observed by MacKinnon, who identified rain shelters, sun-shades, camouflage, and play as the main contexts in which "roofing" was built. In heavy rain and indirect sunlight, orangs often simply piled loose branches and leaves over their heads. The function suggested here could readily explain the propensity of wild orangs to "play" with objects on their heads.

An early account of an infant orang's penchant for putting things on its head also revealed some aptitude for comprehending anatomy. As Cuvier (1811) noted:

> . . . it conceived an affection for two cats which was sometimes attended with inconvenience: it generally kept one or other under its arm, and at other times it placed them on its head; but as in these various movements the cats were afraid of falling, they seized with their claws the skin of the ourang outang, which patiently endured the pain which it experienced. Twice or thrice indeed it attentively examined their feet, and after discovering their nails, it attempted to remove them, but with its fingers only . . . This desire of placing the cats on its head was displayed on a great many other occasions, and I never was able to divine the cause of it. If some pieces of paper fell into its hands, it raised them to its head, and it did the same with ashes, earth, bones, etc. (pp. 197–198)

Figure 1-7. Facial structure of adult male orang-utan *Lipis*. (T. Maple photo)

In captivity, any novel object is usually immediately placed on the head, and even newborn infants are manipulated in this fashion by their mothers.

Rijksen also noted among Sumatran orangs a propensity to hold above or drape over their heads and back, leaves and branches during rain or periods of extensive sunshine. The following passage is instructive (Rijksen, 1978):

The cover was usually loosely draped over the occupant of a nest but in such a regular fashion that the broken ends of the twigs were pointing inwards-up and the foliage direction outwards-down, thus assuring an effective deflection of the water. The cover was often held with one hand and frequently rearranged during the rain, suggesting that the animal adjusted it to stop leaks. The effectiveness of such covers could be assessed from the fact that the orang utans usually had a completely dry appearance after rain. Also imitations of such covers, made by myself during unexpected rainshowers, proved to be surprisingly effective. (p. 153)

Rijksen contends that the nest heights of Sumatran orangs "correspond closely with those of Bornean orang utans," but he asserts that the preferred height is 13–15 metres (30–34 feet) whereas MacKinnon's (1974) histogram for nesting height indicates that the modal nesting height is from 50–60 feet from the ground, with a second peak at 60–70 feet. Thus, if one were to recommend an optimal height for climbing structures in captivity, it would be necessary to choose between two very discrepant estimates.* As MacKinnon has stated:

The height at which nests are made in trees seems to be determined by branch structure. Where several small branches occur close together they can be bent in to form a strong springy platform. Animals appear to get security from height, but there must be sufficient branch support against wind-sway. (1974, p. 48)

RELATIONS AMONG ALIEN SPECIES

There are numerous reports of captive orang-utans which became friendly to dogs, cats, and other animals. Many of these reports revealed the innate curiosity of orangs in their investigations and manipulations of the respective taxa. The propensity of captive orangs

* In a personal communication, Rijksen suggested to me that the height of trees in the wild seems to be often overestimated by observers. For orang-utans, Rijksen believes that it is not height which is preferred but rather the *structure* of the forest. Rijksen therefore recommends captive climbing structures which are moveable, stretchable, and challenging, as is the case in the natural habitat.

to play with an alien species of ape is described in Chapter 2. In the wild, the research conducted by both MacKinnon and Rijksen provides a full description of the orang's natural interactive tendencies.

In MacKinnon's 1974 publication, he reported that orangs were tolerant of other species of animals, but they only interacted with them when they were noisy or threatening. However, humans and dogs were avoided and often vigorously threatened. In fact, orangs have been known to kill dogs.

In Rijksen's field study of Sumatran orangs, he unintentionally exposed some rehabilitants to a caged clouded leopard. An especially interesting response was exhibited by one female as follows:

> The female *Yoko*, who was still recovering from wounds inflicted by the predator two months earlier, even tried to break into the cat's double fenced enclosure. When we withheld her from forcing the lock on the door, she poked with long sticks through the wire, arousing the clouded leopard into defensive reactions. When the cat struck out at the pole, she jumped back and immediately fled high up the tree. (p. 102)

Rijksen, agreeing with Wallace (1869), observed that the adult female orang was especially prone to aggressive display toward humans, while adult males generally hid themselves or simply sat still while staring at the disturbance.

Snakes were generally avoided, but lizards were occasionally the subject of cautious investigation. Of the other animals encountered, the most interesting interactions were with other nonhuman primates. For the most part, interactions with long-tailed (*Macaca fasicularis*) and pig-tailed (*Macaca nemestrina*) macaques, abundant in Sumatra, were peaceful. In fact, one long-tailed male, that fed near the rehabilitation station, was quite friendly to several of the rehabilitant orangs. On several occasions, mutual grooming was observed between two of the young orangs and the friendly long tail. Several agonistic episodes between rehabilitant orangs and pig tailed macaques were reported when competing for provisioned bananas. This fighting may be compared to the induced aggression between baboons and chimpanzees at the Gombe Stream Research Center in Tanzania (cf. Wrangham, 1974).

To summarize, wild orangs are probably too busy making a living to bother with alien species. However, those with leisure time, e.g. rehabilitants, seem to take a greater interest in their surroundings.

In captivity, orangs have occasionally been successfully caged with other species, as at the Baltimore Zoo where recently an adult male successfully lived and occasionally played with chimpanzees. It should be possible to create a natural enclosure for orang-utans which would include many of its rain forest neighbors, such as Hornbills, squirrels, bushpigs, and possibly even langurs or long-tailed monkeys. Such an enclosure would be a most interesting exhibit, and a challenging opportunity for a behavioral scientist.

In the preceding pages, I have reviewed the literature concerning dispersal patterns, distribution, taxonomy, morphology, nesting, eating habits, and the orang's relations with other species. This information, although not exhaustive, should serve as a useful context within which the remainder of this text can be viewed.

Figure 1-8. Orang-utan with proboscis monkey. (From G. Krause, *Borneo* portfolio.)

2
General Behavior of Orang-utan

> The Psychology of the Orang-utan has been thoroughly described by scientists from their observations of the sea-urchin.
>
> Will Cuppy, *How to Tell Your Friends From the Apes*, 1931

To identify the variability in terminology, the reader may wish to consult the oft-cited publications by Rijksen and MacKinnon. In both-of these valuable accounts the behavioral repertoire of wild orangs is well described. However, except for a few important areas which require direct comparison, comment, and clarification, I will rely on my own classification scheme as depicted in Table 2-1. This behavioral repertoire represents the behaviors which are commonly observed in captivity. It is an elaborate but not exhaustive list of behaviors which we may consider to be an appropriate *ethogram* for this species. I will discuss this behavioral inventory in terms of its major behavior systems: locomotion, grooming, temperament, and play. Expression and emotion, including vocal behavior will be covered in Chapter 3.

Throughout this book, there are references to specific units of behavior which have been described by both field and lab researchers. I have made a particular effort to compare many of the behavioral units used by MacKinnon and Rijksen. As the reader will see in examining these tables, there are points of disagreement and potential areas of confusion. For many categories which I have utilized the meaning is clear, and there is considerable overlap with MacKinnon, in particular, regarding the specific elements of the behavioral repertoire. In studies of captive orangs, it is possible to record all or part of the repertoire depending on the aims of the study. A more focused repertoire for studies of mother-infant behavior may be examined in Chapter 5 (cf. Table 5-6).

In discussing those portions of the orang ethogram which comprise the repertoire of social behavior, I am reminded of a paragraph from the Yerkes 1929 volume:

> . . . from our search of the literature we have learned next to nothing. We must, therefore, believe that the scientist who undertakes to study sympathetic relations and activities in this ape will have the advantages of a virgin field of inquiry, and one also of peculiar interest and significance for students of social phenomena.

Despite the marked progress in recent years there is much that remains to be discovered.

Table 2-1. Captive orang-utan ethogram.

Locomotive behaviors
WK—walk quadrupedally.
BI—walk bipedally.
BR—brachiate
CL—climb.
SW—"spider walk"; movement with all fours on ceiling.
SG—swings; walking with support of bars.
SD—slide across the floor with arms on floor in front of animal.
CR—crawl.
LM—locomote with contact or support of another animal.
DR—drop; from ceiling or bars.
VR—ventral ride; on another animal's ventrum.
DR—dorsal ride; on another animal's back (or on head).
AR—arm ride; clinging to a moving animal's arm.
LR—leg ride; clinging to another animal's leg while locomoting.
PV—ventral push; pushing animal along floor with belly down.
PD—dorsal push; pushing animal along floor on its back.

Postural behaviors
SI—sit.
SQ—squat.
LA—lay.
ST—stand.
ZZ—sleep.
SS—stand with support.
HN—hang; by arms or legs and arms combined (see DN—dangle).
VH—"V" hang; supported by legs but with body upright.
NH—net hang; hanging by arms and legs horizontally, arms and legs extended.
BH—"box" hang; sloth hang; suspended by arms and legs, arms and legs parallel.
CG—cling; no support provided by other animal.
CA—cradle another animal in arms.
CF—cradle another animal on floor with arm/hand support.
HA—hold another animal in the air.
HF—hold another animal down on the floor or ground.
DN—dangle; suspended by 2 legs, any one appendage, or 1 arm and 1 leg.
PB—place another animal on bars, or chain.
PC—place another animal on ceiling bars.
PF—place another animal on floor.
SN—stand with contact with or support by another animal.

Table 2-1. Captive orang-utan ethogram (continued).

Self-care behaviors
GS—groom self.
SC—scratch self.
PT—pat self on head or back.
ET—eat.
DK—drink; with lips off wall, with hair on arm, with hand cupped, or from spout.
SR—stretch.
RG—regurgitate food.
UR—urinate.
DF—defecate.

Facial expressions
FF—funnel face; full extension of pursed lips.
BT—bare teeth; top teeth exposed.
GM—grimace; corners of mouth pulled back, mouth slightly open; "grin".
OM—open mouth; teeth showing but lips not pulled back.
YN—yawn.
PY—play face; mouth open, lower teeth exposed.

Vocalizations
LC—long call.
SK—Squeak.
KS—kiss-squeak.
KW—kiss-squeak with wrist or back of hand.
TC—tongue gulp or click.

LOCOMOTION

In our research, we encounter animals which are considerably less arboreal than those which inhabit the wild. Captive animals locomote on the ground in a quadrupedal stance, although they occasionally move in a bipedal fashion. When possible, orangs engage in hand-over-hand brachiation and they may also "spider walk" on barred ceilings, using both hands and feet. Orangs which have been deprived of arboreal activity move cautiously in elevated space, but are capable of improvements in locomotive ability with experience. For example, we found that the male *Bukit*, after living in impoverished surroundings for over 15 years, readily responded to an arboreal challenge when the pathway was punctuated with the first adult female that he had encountered in years! Whether an adult orang be wild or captive, experienced or inexperienced, enriched or deprived, male or female, young or old, compared to most other primates, its activity is generally slow and cautious.

In the wild, orangs are almost exclusively arboreal, although Galdikas-Brindamour (1975) reported recently that adult males come to the ground more often than had been previously reported. In general, all of the principal field workers are in agreement on the modes of locomotion for this species (cf. Davenport, 1967; MacKinnon, 1974; Rijksen, 1978).

MacKinnon's (1974) presentation of locomotion information is especially impressive, and I enthusiastically refer the reader to his paper for further details. Briefly, MacKinnon identified three types of locomotion on the ground: bipedal walking, quadrupedal "crutching," and quadrupedal diagonal couplets. Apparently, bipedal walking is uncommon in wild orangs but is frequently seen among rehabilitants. This characteristic walk requires a shuffling gait with the arms outstretched and sometimes held clasped above the head, apparently for balance. Old drawings of orang-utans often depicted them in a bipedal stance and carrying a stick. These pictures are good evidence that orangs were long ago being raised by humans as pets. In crutch walking, an orang's long arms stiffly touch the ground as their legs swing through, propelling them forward (as if their arms were crutches). The diagonal couplet form of locomotion is complicated to describe, but it is the characteristic form of quadrupedal locomotion for the genus *Pongo*. In this form of movement, orangs move their opposing arm and leg (i.e. left arm, right leg) forward, and follow by moving forward their opposite appendages (right arm, left leg). Although his figures are more instructive, MacKinnon describes it as follows:

> Typically one arm and the opposite leg are placed laterally to form a bridge through which the other arm and leg are swung medially. (p. 38)

GROOMING BEHAVIOR*

In most primate species, grooming serves a significant social function, as well as a hygienic function (cf. Hutchins and Barash, 1976), and resting time is often spent grooming. Such behavior is important to socialization, affiliation, copulation, and dominance. It has been

* This portion of the manuscript follows closely Zucker, Stine, Hoff, Nadler, and Maple, in preparation.

suggested that grooming reduces tension (Terry, 1970), maintains the social group (Zuckerman, 1932), expresses social bonds (Washburn and DeVore, 1961), promotes group cohesiveness (Lindburg, 1973), and allows for the development of interest, sympathy, and cooperation (Yerkes, 1943). Furthermore, Marler (1965) reported a positive relationship between rigidity of dominance relationships and the amount of grooming. The more rigid the social relationships, the more grooming observed. These studies of grooming, however, utilized mainly monkeys. To supplement these findings, we shall examine here the grooming behavior of the orang-utan. In 1933, Robert Yerkes wrote:

> Whereas grooming as social behavior is very frequent in chimpanzee and I have observed it hundreds of times, in orang-utan I have never seen other than casual scratching, mouthing, or picking of a companion, and in gorilla not even this approach to social grooming. My tentative conclusion, from personal experience, checked against the almost negligible contribution of the scientific literature and the oral testimony of several other observers, is that this behavior pattern probably is much more highly developed and more often exhibited as social response in chimpanzee than in any other existing great ape (pp. 9–10).

Since this passage was written, more information has accumulated (cf. Schaller, 1963; Galdikas, 1979; MacKinnon, 1974, 1979). In MacKinnon's reports the orang's grooming was described as both "careful" (1974, p. 37), and "not very thorough" (1974, p. 50). Obviously, more systematic study of orang-utan grooming is required.

In looking at instances of orang-utan grooming which occurred during our studies, data from several different projects were combined, and hence, several different data collection techniques were used. For example, data were obtained from both continuous-living dyads, and pairs brought together for timed-mating sessions. Grooming data on continuous-living animals were collected as part of longitudinal studies of 14 orang-utans at both the Yerkes Primate Center, and at Atlanta's Grant Park Zoo. Timed-mating studies were conducted at the Yerkes Primate Center by Ron Nadler (1976). The latter studies used a check sheet data collecting system providing frequency data only, whereas longitudinal studies provided durations as well as frequencies. These data were first examined in the comparative report of Zucker *et al* (1977, and in preparation). Grooming was defined in our studies as the visual and tactual examination of skin, hair, or nails.

Mode of Grooming

Orang-utans vary in those parts of their bodies used in grooming. They use lips, tongues, fingers, thumbs, backs of their hands, and occasionally their feet and toes for this purpose. Orangs typically, however, move one stiff finger through the hair in *one* direction. In self-grooming, orangs often use only their lips to flip through their hair. When scratching themselves, orangs again use a stiff motion with either an arm, a hand, or fingers, and typically scratch only in one direction.

Areas Groomed

In self grooming, the areas attended are those that the animals can both see and reach. These areas are arms, legs, chest, throat sac and cheek pads. The latter can be pulled into the mature male orang's field of vision. Orangs are highly flexible with strong, prehensile lips (cf. ch. 1), and consequently can reach almost any part of their own bodies. Social grooming is liable to occur anywhere on the other animal's body, but particularly the upper body (e.g. shoulders, head, chest, neck, and arms).

Timed-mating Tests

During Nadler's timed-mating tests, both self- and social-grooming data were collected. Table 2-2 presents the social-grooming data from these dyads of adults, and Table 2-3 summarizes the self-grooming that occurred during these tests.

Table 2-2. Social grooming.

Species	Sex	# of Tests	Occurences of Grooming	Freq. (Grooming/Test)	% of Type of Grooming
Orang-utan	M	99	9	.09	30
	F	99	20	.20	70

Table 2-3. Self-grooming.

Species	Sex	# of Tests	Occurences of Grooming	Freq. (Grooming/Test)	% of Type of Grooming
Orang-utan	M	99	60	.61	60
	F	99	51	.52	40

Under these testing conditions, female orang-utans did far more social grooming than did males. However, the total number of instances of social grooming was rather low. Males, however, engaged more in self-grooming than did the females.

Continuous Living Dyads

Seven pairs of adult male-female orang-utan dyads were observed for a total of approximately 650 hours, with all behaviors being recorded. Five pairs were observed at the Yerkes Regional Primate Research Center, and two were studied at Atlanta's Grant Park Zoo (animals on loan from the YRPRC).* As in the timed-mating studies, self-grooming was more frequent than social grooming (297 vs. 114 instances, respectively). Females, however, engaged in more grooming of *both* kinds, which differs from the timed-mating condition where females social groomed more than males (see Table 2-4). The distribution of social grooming (percentage by each sex) was approximately equal to the timed-mating condition.

Table 2-4. Orang-utan continuous living dyads.

Type of Grooming	Sex	Occurences of Grooming	% Type of Grooming
Social	M	26	23
	F	88	77
Self	M	79	27
	F	218	73

The durations of grooming bouts also differed for the two sexes. Social grooming by females lasted an average of 2 minutes, 34 seconds (range: 4 seconds to 45 minutes), whereas male social grooming

* This research was conducted primarily by M. Beth Dennon (1977) in partial fulfillment of the requirements of the Senior Honors Program at Emory University.

averaged 40 seconds (range: 10 seconds to 2 minutes). For self-grooming, however, the mean duration for males was 1 minute, 20 seconds (range: 5 seconds to 4 minutes, 45 seconds), which was greater than the females' mean duration (48 seconds; range: 5 seconds to 11 minutes, 23 seconds). Although females engaged in more self-grooming under continuous living conditions, the males typically self-groomed for longer periods of time (duration data were not collected during timed-mating tests).

Mother-Offspring Grooming

Donald Sade (1965) has written that adult rhesus monkeys had to grab infants or yearlings, hold them down, and groom them as they tried to escape. This description also applies to captive orang-utans. Young animals capable of locomoting on their own resist grooming attempts by their mothers. Infants, however, are subject to considerable maternal attention. Mothers carefully attend to an infant's hands, feet, and genitalia. Male infants have their genitals groomed both orally and by hand (at the time of our studies there were no infant female orang-utans available to determine sex differences in maternal grooming). Mother orang-utans also trim the fingernails and toenails of infants with their teeth.

According to Jantschke's european zoo observations, the bodies of young animals are groomed with the lips. Seldom did the orangs that he observed use their fingers for this purpose. Particular care was given to the head and ano-genital region, and in the early weeks of life, mothers were especially careful to remove urine and dung from the infants' fur.

Grooming and the Female Sexual Cycle

Cyclic increases in social grooming by females have been observed in both the timed-mating and continuous-living conditions. In timed-mating studies, those females that groomed males at all groomed more during the middle third of their menstrual cycles. Additionally, in one zoo female, increases in grooming (and necessarily, proximity) were observed in cycles approximately 30 days in length, lasting from 4–6 days. Although no hormonal data were collected, it is believed that these peaks were correlated with the onset of ovulation. (cf. Maple, Zucker, and Dennon, 1979).

Special Cases

Two other grooming phenomena require mention. The first is simultaneous social grooming, where animals groom each other at the same time. This has been observed once in orang-utans. A male and female were both lying on the floor when the female began grooming the male's leg, whereby he began grooming her arm. In this case, there may have been some degree of social facilitation. Simultaneous self-grooming is also possible. One male orang groomed his right arm with his left hand, and also groomed his left arm with his lips. Another case involved one animal's interest in another's self-grooming. Female orang-utans show a great deal of interest in a male's grooming of his throat sac, often coming across the cage to closely investigate visually a male's throat sac. It should be noted, however, that many of our male orang-utans undergo periodical throat sac operations because of infection, and the females may learn to anticipate wound-cleaning opportunities.

Although the direct comparability of grooming data across anthropoid species is not possible due to the different techniques employed and the habitat and population characteristics available, our research does provide some notion of orang-utan grooming patterns. With respect to social grooming, there is a great deal of within-species variability; some animals groom, and others do not. MacKinnon (1974) suggested that grooming (self-grooming) occurs in the night nest, where the animals are unobservable. Similarly, grooming in captivity may be an early morning or late evening activity.

The finding that female orangs social groom more often than males is consistent with results from studies of many monkey species (Bernstein, 1970; Bolwig, 1959; Lindburg, 1973; Oki and Maeda, 1973; Rosenblum, Kaufman, and Stynes, 1966; for a review of sex differences in grooming, see Mitchell and Tokunaga, 1976 or Mitchell, 1979). In trying to account for the low *rates* of social grooming in orang-utans, several variables may be considered. However, when compared with apes that apparently exhibit more social grooming (*Pan troglodytes*) these variables cannot account for the differences:

1. Arboreal vs. Terrestrial ways of life: Both chimps and gorillas are more terrestrial than orangs which are overwhelmingly arboreal.

2. Size of social group and its organization: There is no single pattern of social organization among the apes.
3. Hair density or hair length: Gorillas possess dense short hair, whereas orangs and chimps have longer, less dense pelages (Napier and Napier, 1967).

It is possible that a single factor could account for these grooming differences among the apes but, as yet, it remains unknown. However, given the complexity and variability of other primate behaviors, it is highly likely that the determinants of grooming behaviors are similarly complex and variable.

PLAY BEHAVIOR*

While there exist many detailed accounts of primate play, few studies have concerned the great apes. Of these few, the vast majority are studies of chimpanzees (*Pan troglodytes*). In the wild (van Lawick-Goodall, 1968) and in captivity (Savage and Malick, 1976), the chimpanzee exhibits play from infancy through adulthood, and as both captive and field studies indicate, play is an important factor in the normal social development of young apes. According to Loizos (1967) and Groos (1898), the animal that plays (practices) will become more proficient in its behaviors and will therefore hold a selective advantage over less "expert" conspecifics. Of the field reports concerning ape play, Goodall's (1965) is the most complete for chimps. Schaller's (1963) work on gorillas provides useful descriptions for comparison. However, field descriptions of orang-utan play (cf. Harrisson, 1962; MacKinnon, 1974) lack sufficient detail to be useful.

Typical of great ape play studies in captivity is the work of Jacobsen, Jacobsen, and Yoshioka (1932) in which social play was broken down into play-threatening/attacking, and swaggering postures. Other behaviors such as exploration, manipulation, and bodily acrobatics were seen as potentially social. Similarly, Bingham (1927) described chimpanzee play as composed of hand extension, grasping and tumbling, tickling, and chasing. From this report, some excellent notes on forms of parental play are available for study.

* This material is revised and first appeared in Maple and Zucker, 1978.

In our research we have been concerned with the acquisition of data on social behavior and social development in *Pan* (Clifton, 1976; Southworth and Maple, in prep.), *Pongo* (cf. Maple, 1979; Zucker, Mitchell, and Maple, 1978), and *Gorilla* (cf. Hoff, Nadler, and Maple, 1977; Wilson, Maple, Nadler, Hoff and Zucker, 1977). In this chapter the basic components of play behavior for captive orang-utans can be described. For comparison, I will also refer to data acquired on young chimpanzees in their interactions with orang-utan peers. Most writers have agreed that play should be divided into two categories, *social* and *non-social.* Here I will focus on the former, although I will describe the latter as it occurs in certain social or potentially social situations. Non-social play will be described in the context of object manipulation in Chapter 6.

Methods and Procedures

The subjects of our play studies are described in Table 2-5. In all, we have studied 20 orang-utans at play (ranging from infancy to 21 years of age). Our general procedures consisted of daily or three times weekly observation sessions of one to three hours in duration.

Table 2-5. Subject information.

Name	Sex	Age*	Location	Study	Parentage and Rearing
Orang-utans					
Lipis	M	20	Zoo	Sociosexual behavior/ paternal play	Feral born
Bukit	M	20	Zoo	Sociosexual /proceptivity	Feral born
Sampit	M	21	YRPRC	Sociosexual	Feral born
Padang	M	18	YRPRC	Sociosexual	Feral born
Durian	M	18	YRPRC	Sociosexual	Feral born
Bagan	M	19	YRPRC	Sociosexual /proceptivity	Feral born
Dyak	M	19	YRPRC	Sociosexual	Feral born
Sungei	F	21	Zoo	Sociosexual/Infant Development	Feral born
Sibu	F	21	Zoo	Sociosexual/proceptivity	Feral born
Lada	F	19	YRPRC	Sociosexual	Feral born
Paddi	F	18	YRPRC	Sociosexual	Feral born
Ini	F	18	YRPRC	Sociosexual	Feral born
Lunak	M	3.5–5.5	Zoo	Sexual development/ paternal play	Sibu X Lipis

Table 2-5. Subject information (continued).

Name	Sex	Age*	Location	Study	Parentage and Rearing
Merah	M	Birth-1	Zoo	Infant development	Sungei X Lipis
Kanting	M	8–8.5	YRPRC	Sexual development	Jowata X Sampit; Nursery-reared
Loklok	M	7.5–8	YRPRC	Sexual development	Data X Tuan; Nursery-reared
Biji	F	7	YRPRC	Sexual development	Tupa X Dyak; Nursery-reared
Ayer	M	2.5	Play area	Interspecies play	Bali X Tuan; Nursery-reared
Patpat	M	3.75	Play area	Interspecies play	Sungei X Lipis; Nursery-reared
Teriang	M	3	Play area	Interspecies play	Paddi X Padang; Nursery-reared

*Ages given are ages at the time the study was conducted. For feral born animals, ages given are estimates.

Adult Male-offspring Play in Pongo

Our first findings on great ape play were derived during an early study of social behavior in a captive orang-utan group (Zucker, Mitchell, and Maple, 1978). In this study we observed the interactions of the adult male *Lipis*, his consorts *Sungei* and *Sibu*, and *Sibu's* four year old male offspring *Lunak*. (see also Chapter 6). We were surprised at the high frequency of playful interaction between *Lipis* and the young male. Living at the Grant Park Zoo, these animals had been on loan from Yerkes and were housed together for the duration of our three month study until their breakup due to the injury of *Sibu* by *Lipis*. From this research, we obtained 28 concise descriptions of "paternal" play, 26 from films and 2 from written notes.

In these play bouts, ten principle behavioral components (A–J) were identified (Table 2-6). The most frequent of these was *non-aggressive biting* (mouth fighting), which occurred 39 times, nearly 25% of the total number of behavioral events. The second most common behavior was *hand contact* which occurred 34 times, and appeared at least once in 57% of the interaction sequences. The other frequent components were *dragging/pushing* (24), *following*, and *extremity* or *head investigation* (14 each). In decreasing frequency, the remaining behaviors were *mouth contact* (11), *hand extension* and *hair pulling* (7 each), *falling* (4), and *face stroking* (3).

Table 2-6. Behavioral categories identified.

A. Hand contact: haptic contact with any part of the other animal's body. Includes swatting—brief contact w/continued arm/hand movement.
B. Nonaggressive biting or mouth fighting: contact with other animal's body with open mouth, teeth visible.
C. Following: remaining proximate to the other animal as the other animal moves.
D. Dragging/pushing: movement, in contact w/floor, towards or away from the other animal as a result of impetus applied by other animal. Includes rolling of other animal.
E. Extremity or head investigation: touching of other animal's appendages or head.
F. Mouth contact or oral exploration: contact w/other animal w/closed mouth or lips/-tongue.
G. Hand extension: no contact made w/other animal.
H. Hair pulling: common usage.
I. Falling: movement from a standing position to a sitting or prone position without impetus from another animal.
J. Face stroking: slow, repetitive vertical contact w/face by fingertips of another animal.

In four of the 28 interactions, the initiator could not be clearly determined. However, in 20 of the remaining 24 cases, *Lunak* initiated play. *Lipis* clearly initiated only 4 play bouts. Similarly, *Lunak* terminated 16 of the interactions, while *Lipis* ended contact only twice. The mode of termination was generally withdrawal from proximity, whereby the young male usually returned to his mother.

There were two common positions for these playful interactions; both subjects on the floor face-to-face, and one or both animals hanging from the ceiling bars. On those occasions when only one partner was hanging, the other was lying on the topmost platform. Both of these occasions when only one partner was hanging, the other was lying on the topmost platform. Both of these basic positions occurred with equal frequency, 13 times each (43% of all interactions). In the hanging position *Lunak* hung the most frequently, accounting for 12 of the 13 events. Less frequent positions were with the adult male on the floor and *Lunak* on the lower bench, and both on the floor. As can be seen, social play in this pair reflects, in many instances, the arboreal propensities of the orang-utan (see also adult heterosexual play).

In examining the order of these recorded play components, 9 of the 28 bouts began with *hand contact*, 5 with *hand extension*, and 4 with *head* or *extremity investigation*. Thus, contact by the hand of one or both subjects was the initial mode of interaction in 60% of the playful sequences, not too surprising a finding in light of the tactual skills of great apes. An inspection of the sequences revealed that the most common sequential pair of behaviors was *nonaggressive biting* following *hand contact*. After this pair in frequency was *dragging-pushing* followed by *nonaggressive biting*, and vice versa. Other common pairs were *hand contact* followed by *nonaggressive biting*, and *nonaggressive biting* followed by itself.

Father-infant play in this situation may be characterized as rough-and-tumble. In this fashion, it is different than play between *Lunak* and his mother (*Sibu*). Although *Lunak* played rough with *Sibu*, these interactions were generally unidirectional with *Sibu* tolerating play, but not reciprocating in kind.

Lipis' rather low rate of play initiation suggests that adult male orangs, though vigorous players, may not themselves seek out younger play partners with any great frequency.* I am confident in labeling these interactions as playful, since fearful vocalizations were absent, behavior was reciprocal, neither animal attempted to flee and interactions were initiated and sustained by the smaller animal. In this latter category, the data differ from that of Jantschke (1972) who reported that the more dominant animals initiated play.

Regarding the motor patterns reported in this study for orangs, it is interesting to note that *hand extension* and *hand contact* are also integral parts of play for chimpanzees (cf. van Lawick-Goodall, 1968) and gorillas (Wilson *et al*., 1977). *Hand contact* and *hand extension* may also serve a *metacommunicative* function (cf. Altmann, 1967), indicating that whatever follows will be play rather than aggression.

Peer Play

Although we are just beginning to study peer play in infant and juvenile gorillas and orangs, some preliminary data on conspecific interactions and a detailed description of play behavior between juvenile orangs and chimps can be reported here.

* In a recent study conducted by De Fiebre (1979) considerable play behavior between an adult male and his infant offspring was observed. In these instances, contrary to our observations, the adult male initiated most of the contact. This case illustrates the variability inherent in the repertoire of the species, and also the hazards of relying upon a single case study.

We recently completed the first phase of a study of three juvenile orang-utan males (two nursery- and one mother-reared) and their initial responses to non-maternal heterosexual contact. We had hypothesized that the mother-reared male would emit more appropriate sexual behavior than would the two nursery-reared males. In observations of their behavior together (cf. Zucker *et al.*, 1977), the mother-reared male was the only one to emit pelvic thrusting toward his like-sexed cage-mates. To our surprise, differences in sexual behavior were not apparent in the heterosexual condition as all three animals responded to the six year old (heterosexually naive) female with vigorous play. It is quite possible that orang-utans respond to all peers playfully until they have reached adulthood, or some other threshold of sexual development. While it is too early in this research to say that appropriate sexual behavior can be learned through play, it is clear that captive mothers provide some early input by *mounting* and *thrusting* against their babies at an early age (cf. Maple, 1979). We have observed in these animals evidence of sex differences in rough and tumble components. As is the case with adult pairs, the sub-adult female *Biji* played just as vigorously as did the three males with which she was housed.

Play with Alien Peers

Species differences in the behavior of captive-reared juvenile chimpanzees and orang-utans in conspecific groups were reported by Nadler and Braggio (1974). Our research was designed to extend these findings to a mixed-group situation, complementing my previous work on intertaxa social behavior (cf. Maple, 1974; Maple and Westlund, 1975; Maple and Westlund, 1977). The first objective was to determine the relative tendency of young apes to affiliate with members of an alien species of comparable age. Second, we wished to contribute to the development of our great ape ethograms by describiing the motor patterns emitted during these interactions, looking for species typical modes of interaction. We were surprised to note that the vast majority of interactions between species were playful. The subjects of this study were three juvenile orang-utan males, and three juvenile chimpanzee females which were available at the time of the study (orangs: *Ayer*, *Patpat*, *Teriang*; chimps: *Barbara, Ellie,*

Joice). We acquired our data during two-hour observation periods, during which time we united the animals and recorded behavior through the use of a super-eight movie camera and handwritten notes. One observer filmed play interactions as they occurred (which is actually a sample, since there were several playful interactions occurring at once, and play was essentially continuous), while the other observer recorded frequencies of both interspecific and intraspecific contact during 15-minute time samples. From the films, 27 behavior categories were identified (Table 2-7) from 65 discrete interspecific interactions.

Table 2-7. Behavioral categories in interspecies play. (From Maple and Zucker, 1978.)

1. Grabbing: Rapid movement to hold other animal by putting arms around other or gripping other with hands.
2. Hair grasping/hair pulling: Common usage.
3. Non-aggressive biting: Contact with other's body with open mouth, teeth visible.
4. Climbing up apparatus: Climbing, jumping, and/or swinging on apparatus.
5. Pushing/pulling: Movement to free self from other animal.
6. Reaching toward another animal without contact.
7. Jumping to ground: Jumping or dropping to ground from a climbing structure, fence, or table.
8. Mouth fighting: Reciprocal contact with another's face with an open mouth.
9. Hand contact: Haptic contact with any part of another animal's body.
10. Object stealing: Grabbing or pulling an object away from another animal.
11. Jumping over other: Jumping/swinging around or over another animal.
12. Walks sideways: Takes steps to the side while watching another animal.
13. Hand grappling: Reciprocal grabbing and pulling at the hands of another animal.
14. Jumping/falling on other: Jumping or falling on another animal without impetus supplied by another animal.
15. Dangling: Holding onto a structure with arms or legs hanging unattached to the apparatus.
16. Shoving/pushing: Forceful movement against another animal. Include shoving, running or swinging into other, and pushing.
17. Hitting: Slapping with open hand or hitting with fist.
18. Play thrusting: Slight back and forth motion of body while standing quadrupedally.
19. Jumping on other: Jumping up and down *on* another animal.
20. Swinging around bars: Grasping and swinging around vertical bars in a circular motion.
21. Grooming: Picking at hair of another animal.
22. Jumping near other: Jumping up and down *near* another animal.
23. Walking backwards: Walks backwards dragging hands and arms in front of body.
24. Following: Remaining proximate (arm's length away or less) as the other animal locomotes.
25. Mouth contact: Touching another animal's body with a closed mouth, or with lips and/or tongue.
26. Slapping ground: Slap hands against ground or table.
27. Presents: Stands quadrupedally with ano-genital region towards another animal.

A summary of play contact initiated during the six 15-minute samples is presented in Table 2-8. Of the 216 total contacts, 198 (91.7%) were interspecific, while only 18 (9 chimp-chimp; 9 orang-orang) were intraspecific. Of the 198 interspecific interactions, 128 were initiated by chimpanzees and 70 were initiated by orang-utans ($X^2 = 17.18, p < .001$).

Figure 2-1 shows the frequency for the specific behaviors that began interspecific interactions for 50 such interactions. For 15 interactions, the initiating behavior could not be determined. The most frequent behavior initiating play was *grabbing*, followed by *object stealing, hand contact, hair grasping/pulling, shoving/pushing, hand extension*, and *jumping over other*.

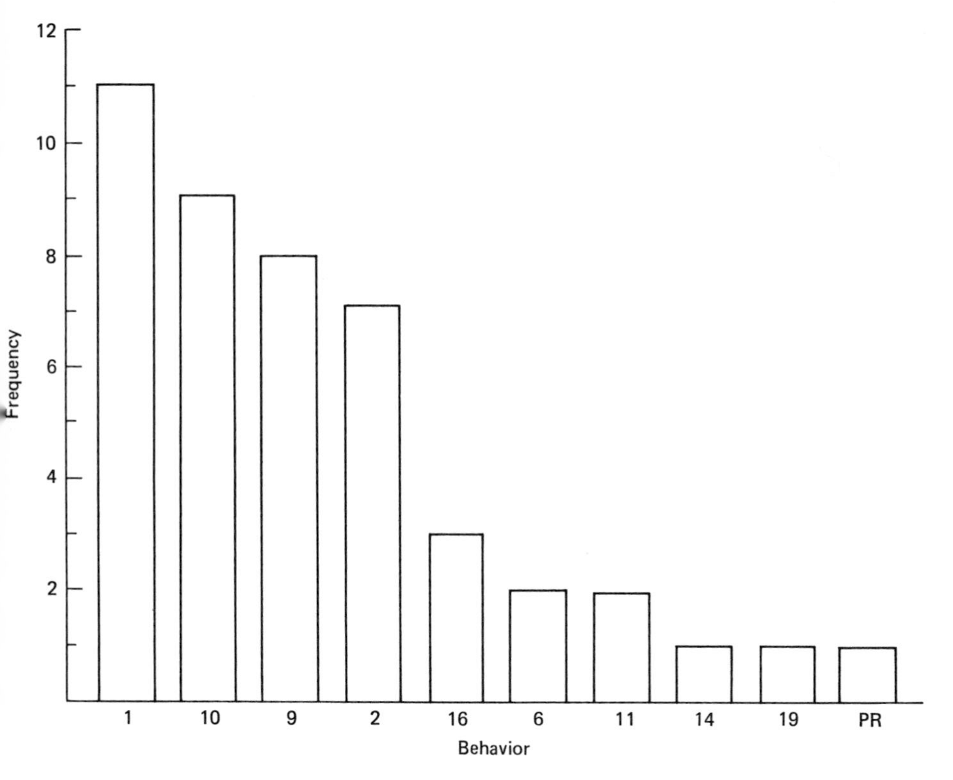

Figure 2-1. Behavioral categories observed in descending order of frequency during interspecies play. Numbers refer to behaviors emitted by both orangs and chimps (see Table 2-7 for corresponding behaviors).

Table 2-8. Interspecific and intraspecific interactions for six observation periods, broken down by species, initiator, and recipient.

	Observation Period						
Initiator—recipient	**1**	**2**	**3**	**4**	**5**	**6**	**Total**
Interspecies							
Chimp—Orang	21	26	40	11	13	17	128
Orang—Chimp	18	2	8	9	16	17	70
							198
Intraspecies							
Chimp—Chimp	3	1	2	0	2	1	9
Orang—Orang	1	0	1	0	3	4	9
							18

Orangs and chimps exhibited significantly different behaviors within the play sequences (see Table 2-9, $X^2 = 240.96$, $p < .001$). Table 2–10 presents the total frequency of each behavior by species during the 65 interactions. Of the 13 most frequent behaviors, chimpanzees emitted more *climbing, pushing/pulling, object stealing, jumping to ground* ($U = O$, $p < .200$).* Two categories necessitate reciprocity, *mouth fighting* and *hand grappling*, and there were no observed species differences in the incidence of these behaviors. One behavior, *presents*, involved both a species and a sex difference, but this behavior rarely occurred. The number of behaviors per interaction ranged from 1 to 66 (Mean = 10.7; Median = 8).

The high degree of interspecific play which occurred during this study is remarkable, as ample opportunity existed for conspecific play. It is not entirely surprising that young chimpanzees and orang-utans should interact. For all their differences, they are also quite similar in size, structure, and in many of their facial expressions. However, these similarities are not a prerequisite for interaction, since very diverse primate species have been known to interact in an affiliative manner (Maple, 1974).

Differences in the mode of play behavior observed in this study reflect fundamental species differences. Chimpanzees are fundamentally terrestrial animals, and live in communities composed of a number of small groups (van Lawick-Goodall, 1968). Orang-utans, as we have seen are principally arboreal apes, living either as solitary

* The remaining behaviors were too infrequent to compare.

Table 2-9. Total frequencies of each behavior for each subject in the interspecies play study. Refer to Table 2-7 for the behaviors that correspond to the numbers.

	Behavior																											
Subject	1	2	3	4	5	6	7	8	9	10	11	12	13	14	15	16	17	18	19	20	21	22	23	24	25	26	27	Total
Barbara	11	14	5	33	21	1	21	22	7	13	12	0	5	3	8	5	7	4	10	3	1	2	0	4	0	0	1	213
Ellie	11	3	5	17	17	3	11	11	8	5	5	2	10	3	5	6	4	5	0	1	1	1	2	6	0	3	1	146
Joice	0	0	0	1	1	0	0	0	2	0	0	0	0	1	0	0	0	0	0	0	0	0	0	0	0	0	0	5
Ayer	21	14	26	7	1	6	1	9	0	0	0	0	5	0	0	0	0	0	0	2	0	0	0	3	0	1	0	96
Patpat	40	14	15	6	3	12	0	25	8	2	0	0	9	4	0	1	0	0	0	2	0	0	0	1	1	0	0	143
Teriang	19	13	8	13	1	13	4	4	9	3	1	0	1	0	0	1	0	0	0	0	2	0	0	6	2	0	0	100

Table 2-10. Total frequencies of each behavior for each species in the interspecies play study. Refer to Table 2-7 for the behaviors that correspond to the numbers.

	Behavior																											
Species	1	2	3	4	5	6	7	8	9	10	11	12	13	14	15	16	17	18	19	20	21	22	23	24	25	26	27	Total
Chimps	22	17	10	51	39	4	32	33	17	18	17	2	15	7	13	11	11	9	10	4	2	3	2	10	0	3	2	364
Orangs	80	41	49	26	5	31	5	38	17	5	1	0	15	4	0	2	0	0	0	4	2	0	0	10	3	1	0	339
Total	102	58	59	77	44	35	37	71	34	23	18	2	30	11	13	13	11	9	10	8	4	3	2	20	3	4	2	703

animals (adult males) or in groups of two or three, as in the case of females with their offspring (Davenport, 1967; MacKinnon, 1974; Rodman, 1973). These differences in use of vertical space are retained in captive-born animals (Nadler and Braggio, 1974). Behaviorally, chimps are very active, while orangs are lethargic, slow-moving, and deliberate (Yerkes and Yerkes, 1929; Davenport, 1967).

Because of these activity differences, it was expected that chimpanzees would engage in more active types of behaviors, and would initiate more interactions because of their greater activity and gregariousness. The results supported these predictions, although both species initiated a large number of interactions.

One explanation for the high frequency of interspecific interactions relative to intraspecific interactions is that both chimpanzees and orang-utans responded to each other as novel stimuli. However, this tendency appears to have been remarkably resistant to habituation since it did not wane with time. The novelty of the situation may have been maintained by the fact that the opportunity to interact was limited to only two hours/day. Extended exposure per day may have resulted in a decrease in interactions, and a decrease may also have been observed had the study continued for a longer period with repeated two-hour play sessions.

Alternatively, as our data indiciate, an interaction with an animal different in its mode of play may have been reinforcing in some manner. Another reinforcing property of these play sessions may have been the opportunity to interact with an animal of roughly the same size. With the exception of *Joice*, the heavier animals interacted the most, the lighter animals the least. Nadler and Braggio (1974) also found body weight rather than age to be correlated with the number of interactions in conspecific groups of chimpanzees and orang-utans. We suspect that *Joice's* extremely low interactive rates and behavioral output may be due to brain damage incurred either prenatally or at parturition. *Joice* was a dizygotic twin, delivered by Caesarean section, whose respiration needed to be stimulated with drugs. These birth traumas may be reflected in her social and play behaviors. On the other hand, the orang *Teriang's* high interactive rate can be explained by either the novelty or body weight hypothesis. Of all the animals, *Teriang* was the heaviest, but he was also the most novel of the three orangs. A Bornean orang, he had a darker brown coat color than either of the other two orangs (Sumatrans), as well as more developed cheek pads and throat sac.

Freeman and Alcock (1973) found that the majority of *intraspecific* social play of their young gorillas involved heterosexual interactions. In this study, interspecific interactions were exclusively heterosexual, but the only sexual behaviors that occurred were two *presents* by chimpanzees (*presents* are also chimpanzee submissive social signals). The absence of any other sexual behavior suggests that these *presents* were communicative in nature, rather than sexual.

It is possible that behavioral differentiation between the two species studied was due to the sex of the animals (and we recognize the limitations of our sample), but these differences can be explained more parsimoniously as species differences. More climbing and jumping behaviors were observed in the chimps than in the orangs, as well as more object stealing. The orangs, however, displayed more behaviors which involved greater use of the upper portions of their bodies, and less movement of their legs. It is surprising that they exhibited less climbing.* There was more hair pulling, grabbing, and biting by the orangs than by the chimps. The specific behaviors were often employed by each species to gain an advantage over the other. Chimps utilized their greater agility in play bouts. Often, a chimp climbed or hung on the fence or climbing structures above an orang's head until the orang grabbed for it, then quickly withdrew out of the orang's reach. Also, chimps slowly climbed on the apparatus, and as soon as the orang caught up, the chimp dived through the bars and dropped to the ground. Chimps were also observed to jump from an apparatus, and over an orang by placing their hands on the orang's shoulders and hopping as in "leap frog."

In this study, the orang-utan's strength compensated for the chimpanzee's agility and speed. Orangs often ignored a chimp's presence only to suddenly lunge forward, grab a chimp's arm, leg, or hair, and bite. The orang's strong hold prohibited the chimpanzee from escaping. Several behaviors exhibited by chimpanzees were rare or absent among the orang-utans' behaviors. These included dangling, play-thrusting, walking backwards, walking sideways, jumping on other,

* The climbing apparatus was located in the central portions of the cage. Only one of the orangs spent its time in these inner portions. All three orangs were peripheral relative to the more mobile chimps. This location may have been a factor in this difference.

presenting, and hitting. Orangs exhibited a greater degree of mouth contact than did chimps and repeatedly bit the face, hands, feet, and back of chimpanzees during play. Nadler and Braggio (1974) reported a higher incidence of slapping and hitting behavior for chimpanzees than orang-utans. In this study and our study, chimpanzees spent more time on the ground than did the orang-utans. The arboreal orangs' survival in its natural habitat is dependent upon strong grasping behaviors, making the orang's mouth and lips important structures in interacting with their social and physical environments.*

When orangs moved from one area to another, they usually moved around the periphery of the cage, holding the fence as they walked on the ground, or climbed around on the fence. Other captive animals are known to remain near edges in their environments (cf. Menzel, 1965). For our captive orang-utans, however, their locomotion was highly dependent upon these boundaries, unlike the chimpanzees. In play interactions the orangs were frequently observed to lie on their backs, holding on to a climbing structure with their feet. Even when not interacting, the orangs would keep at least one hand or foot attached to a permanent structure in their enviroment (see also Chapter 4.)

It was not always clear which behavior actually initiated a given play bout, since physical contact was not completely necessary. That is to say, subtle gestures cannot always be detected on film. Play bouts can be started by simply gaining proximity to another animal, and terminated by either moving out of proximity or by discontinuing behavioral output. Criteria must be established by which initiation and termination can be judged.

---

* It is interesting to note that Abel in 1818 clearly described the orang play-style, as differentiated from the chimpanzees in our studies. This description clearly indicates the grasping nature of the orang.

> He would entice them into play by striking them with his hand as they passed, and bounding from them, but allowing them to overtake him and engage in a mock scuffle, in which he used his hands, feet and mouth. If any conjecture could be formed from these frolicks of his mode of attacking an adversary, it would appear to be his first object to throw him down, then to secure him with his hands and feet, and then wound him with his teeth. (Yerkes and Yerkes, 1929, pp. 327-328).

Hornaday (1885) also noted that orangs often bit their partners' fingers and toes during play, a common play behavior as indicated by our studies as well.

> One of his favorite tricks was to seize my hand suddently, draw it to his mouth, and make a feint of giving it a terrible bite. But he always knew that he must bite gently, which is more than can be said of any human infant I every experimented with . . . (Yerkes and Yerkes, 1929, p. 383).

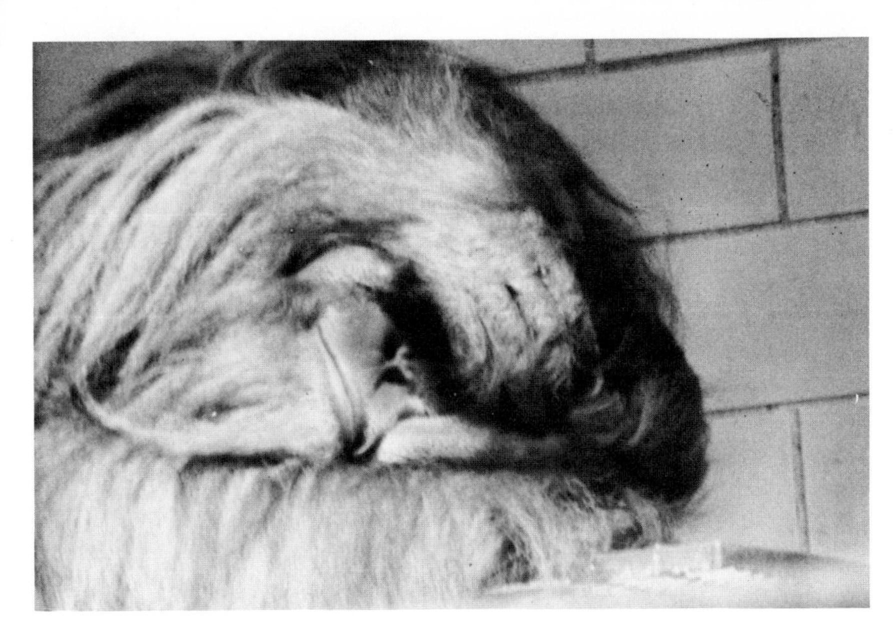

Figure 2-2. Resting posture of captive adult male orang-utan (T. Maple photo).

Figure 2-3. Resting posture of captive adult female (E. Zucker photo). Note position of hand over head.

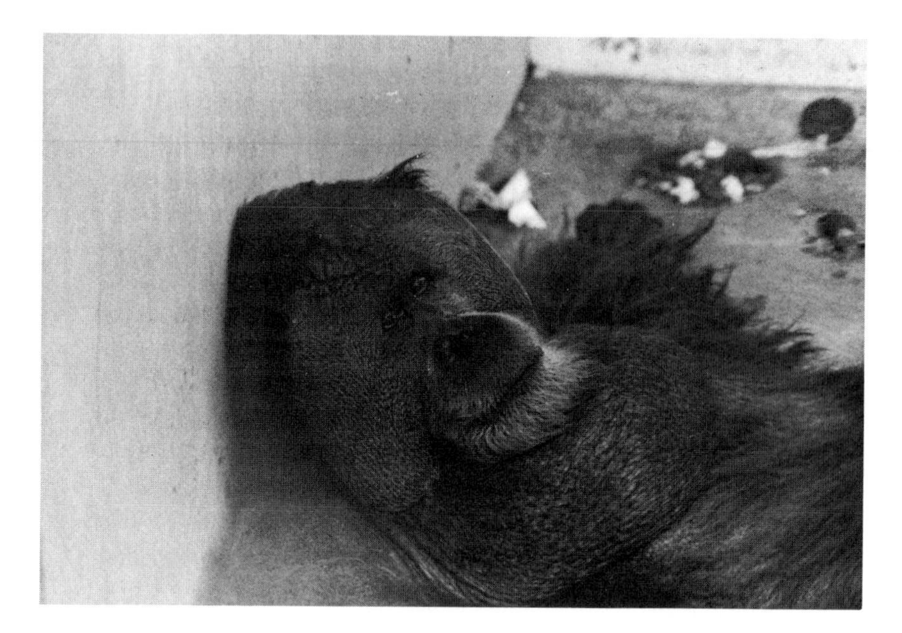

Figure 2-4. Resting posture of adult male *Bagan* in barren captive habitat (E. Zucker photo).

Adult Solitary Play in Pongo

Our studies of adult heterosexual pairs of orang-utans have been carried out in both the Yerkes and Atlanta Zoo environments. Both solitary and social play are evident in these settings, although there are considerable individual differences in the amount of play emitted. In solitary play, we have been concerned with self-motion play (or *peragration* as Mears and Harlow (1975) have called it) rather than object play. While the latter has not been the focus of our work, we have noticed that both orang-utan and gorilla adults play with objects in similar ways. Both species put objects on their heads, slap their hands onto the ground while sitting and holding objects, pound rocks and sticks repeatedly, throw objects into the air, and roll around with objects in the same way that they roll with another animal (often emitting *play faces*). The self-motion play that we have seen in *Pongo* has been observed only in females. In female orangs, self-motion play is composed of *hanging* or *dangling* by the hands or feet, *somersaulting*, rapid *swinging* and *brachiation, spinning, sliding, dropping to the floor, water stomping*, and *pounding* the floor, all of which are typically accompanied by a *play face* and a vocalization which we call *tongue-gulps* (vocal clicks). In the seven adult pairs studied recently at both Yerkes and the Atlanta Zoo, solitary self-motion play was found to be characteristically brief in duration, 1 minute and 12 seconds on the average (range: 10 to 12:50). In three of our seven females, solitary play was especially prominent in their repertoire. In three others solitary play occurred, but infrequently, while in the remaining animal, solitary play was absent.

In agreement with MacKinnon (1974) our studes revealed that adult female solitary play can be followed by copulation. In fact, solitary play may be an indicator of female hormonal status (see Chapter 4), as we have found evidence in one female that it correlates with proceptive sexual behaviors (Maple, Zucker, and Dennon, 1979). Residing in the zoo environment, the adult female *Sibu* emitted play behavior which consisted of cyclic periods of play overlapping the proceptive behaviors of *following, positioning*, and *pelvic-thrusting* against the male. During the course of our research into orang sexual behavior, *Sibu* exhibited gradual increases in solitary play spanning the days 12–32, 46–55, and 73–94.

Solitary play was not observed in this animal after the 94th day which coincided with the onset of the female's pregnancy.

The male *Bukit* showed interest in these activity periods and frequently approached the female during her solitary play bouts. As a result of this interest, social play followed on three occasions. In one case, *Bukit* chased *Sibu* after a solitary play bout and engaged in ventro-ventral copulation.

The female *Lada* also exhibited a cyclic trend in solitary play while paired with *Bagan* at Yerkes. These play peaks occurred during the days 21–44, 57–70, and 79–94. Coincident with these play peaks were increases in social grooming by this female, although the male did not appear to take interest in these activities.

Adult Social Play in Pongo

In our observations of captive orangs we have observed frequent social play in two pairs, less frequent play in four pairs, and no social play in one pair. During the course of these observations, 93 bouts of social play were recorded, and all but 7 were initiated by the females. Social play was initiated most commonly by the female dangling by her feet above the male (cf. also adult male-offspring play) and extending her hand to him. While in this position females typically hit or slapped the male. The male's response was generally to pull the female to the floor after which a rough and tumble play bout occurred. An alternative mode of initiation occurred with the female in a sitting position. In these instances, the female rocked into contact with the male with her head (which we call *head butting*). When males initiated play the initiating behaviors were the same as those of females; *hand extension, head butting*, and *dangling-hitting*.

During the course of adult heterosexual play bouts, we identified a number of principal components, such as *wrestling, rolling other, slapping, hitting, mouth fighting, biting hands/feet*, and *hair pulling*. These behaviors were consistent with other descriptions of orang-utan social play (cf. Abel, 1818, cited in Yerkes and Yerkes, 1929; MacKinnon, 1974; Yerkes and Yerkes, 1929; Zucker, Mitchell, and Maple, 1978). It is interesting to note that in our study, thus far, copulation has never been observed to follow a bout of *social* play.

The average duration of social play bouts in our study of adults was 3 minutes, 19 seconds. Scores ranged from 9 seconds to 23 minutes and 25 seconds. To end a bout, females withdrew during play more frequently than did the males. In 82 play bouts where withdrawal was recorded, 60 were female withdrawals (73%) Female-initiated social play was cyclic for the female *Sibu*, who played at highest rates on days 16, 46, 51, and days 73–93. As in the case of solitary play, this activity was coincident with proceptive behaviors as previously described. During her pregnancy, female-initiated social play was not observed. Although *Lada's* play with *Bagan* was not cyclic, she emitted a high frequency of social play after day 56 until the end of the study.

Play Signals

Since it occurs so often during chimpanzee social play, the wide-open mouth with covered or partially covered teeth (as described by van Hooff, 1962) has become known as the "play-face." This expression has also been observed in gorillas and orang-utans. However, the presumed function of the play-face, in our view, has not been precisely determined (indeed, it may not be possible to do so). For example, in Loizos' review (1967), she described a modified chimpanzee play-face as observed by Goodall at the Gombe Research Station in Tanzania:

> . . . it is possible that this represents a high intensity of play . . . it is also possible that in certain socially ambiguous situations the increased baring of the teeth (and particularly of the upper row) in the play-face represents the introduction of an element of appeasement since the facial expression now bears more resemblance to the grin face, which can have an appeasing function (Loizos, p. 205)

The play face itself has also been described as an appeasement gesture, informing the other aninal that its approach is not an aggressive one. In this sense, the play-face may fulfill the definition of *metacommunication* (cf. Altmann, 1967).

In our research we have found the play-face to be an unreliable indicator of play in orang-utans in that it may or may not precede a vigorous play-bout. Moreover, because of the certainty of play-biting, the play-face is difficult to distinguish from biting when the animals are in contact. To make matters worse, the play-face occurs as a transition from vocalizations to grimaces, and is so transient as to be useless during intensive play-contact. Thus, we have found the play-face to be of very limited use, in that its predictive value has not yet been empirically verified. That it often occurs in connection with play, however, cannot be denied. Its utility and social function need to be further explored.

Play in the Field

Agreeing with the rest of us (cf. Maple and Zucker, 1978), Rijksen found play equally difficult to define in the field. His observations of play indicated that it was composed of mutual and reciprocal contact and was often preceded by or included *meta-communicative* elements. Recognized in the play repertoire were the following elements:

play approach	self-decorate
grasp	ignore
gnaw wrestle	gymnastics
hitting	relaxed open mouth
mouth to mouth	ahh vocalization
nest building	

The elements which Rijksen suggested as preceding *meta-communication* were relaxed open mouth, ahh, loutish approach, self-decorate, open-mouth, bared-teeth expression, and dangling by feet.

Immature orangs observed by Rijksen played more frequently than adults. Male-male play accounted for 44% of all male play whereas female-female play accounted for only 10% of a female's play time. Thus, males were the more sought after play partners by both sexes. In addition, males played rougher with other males than with females as indicated by an increased tendency to *gnaw-wrestle* with other males. Invitations to play with the opposite sex was more a female than male behavior (34% to 11%). Male-male play bouts lasted longer than did female-female play bouts (5.4 min. to 2.6 min.).

Rijksen also found that the frequency and duration of play varied considerably among individuals. However, he also suggested that dark haired, long fingered types were less playful than the light haired, short fingered type. (see also Chapter 1).

Rijksen never saw fully adult wild males engaged in play, but mothers occasionally played with both infant and older offspring. None of these findings contradict the results of our research on captive orangs. As we are able to confirm, males play rougher than females, and fully adult males rarely play. Rijksen's research suggests that orangs will be orangs whether in captivity or in the wild.

Play as Preparation for Coitus

The links between early social experience and subsequent sexual competence in monkeys and apes have been the subject of many experimental, developmental, and physiological studies (cf. Harlow, 1971; Mason, 1963; Rogers and Davenport, 1969). The verification of such links in the great ape family has been difficult since the latency from birth to sexual maturation is nearly a decade. Harlow

Figure 2-5. Young apes at play in zoo nursery (San Diego Zoo photo). Note "solitary" attitude of orang-utan.

Figure 2-6. Young gorilla and orang at play (San Diego Zoo photo). Note fear grimace of orang-utan.

and his associates (Harlow, 1962; Senko, 1966) reported the occurrence of rudimentary adult sexual behaviors in play bouts of laboratory-reared infant and juvenile rhesus monkeys, including behaviors such as mounting and thrusting. Some of these behaviors, however, have been identified as species-typical social signals in addition to having acknowledged sexual functions. From these studies, it has been suggested that a function of peer play is to facilitate the acquisition of successful reproductive behaviors.

The orang-utan pattern of copulation is characterized by the male's forceful pursuit of the female. This sequence of events has acquired the unfortunate label of "rape" (cf. MacKinnon, 1974; Nadler, 1977). A typical copulation is composed of a precopulatory chase, followed by intromission, thrusting, and subsequent ejaculation. During such forceful copulation, the female struggles to break contact, and as a result, the partners repeatedly assume various postures (cf. Chapter 4). Under these conditions, the interaction is often of considerable duration (cf. Nadler, 1977; Maple, 1979). However, the behaviors necessary to fulfill the male or female's roles in copulation are present in the repertoires of both sexes since females apparently possess the capability to initiate copulation, i.e. to emit proceptive behaviors (see Maple et al., 1979, Chapter 4) which closely resembles the typical behaviors of adult male orang-utans.

The aim of one of our recent studies (Zucker, Puleo, and Maple, in prep.) was to provide evidence that in captive orang-utans sexual behaviors are developed through play. The data reported in this paper were based on an analysis of the component behaviors of play rather than on global categories such as rough-and-tumble, approach-avoidance, or contact play, which permitted a more detailed description.

The subjects of the study were seven immature orang-utans. *Jinjing* and *Anak* were housed together with *Jinjing's* 20-year-old mother and his 10-year-old nulliparous female sibling. *Anak* was nursery-reared for his first 1.5 years of life. For part of the study, the adolescents were housed sexually segregated, but were integrated during the latter period of the study in two separate, nonadjacent cages as the result of a colony management rearrangement. None of the adolescents had any extensive prior heterosexual experience. *Lunak* was mother-reared at the Atlanta Zoological Park for his first four years of life (Zucker, *et al.*, 1978), whereas all other adolescents had been nursery-reared, and had social contact only with male peers.

Anak and *Jinjing* were observed for 40 sessions over a 6-month period, totalling 36 hours of observations. Their play behavior was recorded in the course of a study focusing on *Anak's* social rehabilitation following nursery-rearing (Puleo *et al.*, in press; Chapter 5). The three adolescent males were each paired with the female *Biji* for four one-hour sessions, with the order of pairing randomized. None of the males was tested on more than two consecutive days. During the sessions, the two males not being tested on that particular day were confined to the indoor portion of the cage, and *Biji* was introduced to the test male in the outdoor compartment. Pairing of the males with *Bunga* was intended to be done in the same manner, but strong peer attachments prohibited the daily transfer of the subjects without undue stress. Because of these transfer difficulties, each male was paired with *Bunga* for only one one-hour session. Furthermore, *Bunga* could not be moved as easily as *Biji,* which also prohibited the testing of the peer play of the two females together. Data collection began immediately after introduction to minimize habituation and to maximize the probability of play.

Behaviors were recorded sequentially using our typical two-letter code, noting the initiator and recipient of all acts. The duration of each play bout was recorded, as was the duration of each behavior whenever possible. Chases, for example, could easily be timed by noting the starting and stopping times from a continuously running stopwatch, although short-duration acts such as *hit* could not be accurately measured nor recorded.

The data reported here represented 51 total hours of dyadic play: 36 hours for the male juveniles *Jinjing* and *Anak*; 5 hours for each of the three adolescent males; 12 hours for the female *Biji*; and 3 hours for the female *Bunga*. Since the amounts of observation time varied across subjects, raw frequencies of component behaviors were converted into relative rates (occurrences/hour of observation/individual animal), as well as relative rates for each age/sex class (total occurrences/total hours for the age and/or sex class). The small sample size precluded statistical analysis, but relative rates permitted comparisons of the respective play behaviors.

The single most frequently occurring behavior was wrestling, exclusive of other behaviors occurring simultaneously. The overall mode of play was "rough-and-tumble," and this contention is supported by the frequencies of the component behaviors, though the distribution of these other behaviors among animals provided more information about the nature of the play bouts. Comparison of the relative rates revealed both age and sex differences in several component behaviors. Table 2-11 presents these behaviors in terms of the mean relative rates for each age/sex class: adolescent males, adolescent females, and juvenile males. The most striking sex differences were in the relative rates of *chase* and *grab*, with *bite other, pull*, and *touch* also predominantly male-emitted behaviors. For these behaviors, except *touch*, there were developmental differences as well, with juvenile (2-year-old males) emitting these behaviors at a rate intermediate to that of the adolescent males and the adolescent females. *Roll* was predominantly a female behavior, whereas *escape* and *forward roll* were exclusively female-emitted. Though *forward roll* was a specific behavior, it too served to break contact with a male, and could then be functionally labeled an escape behavior.

Table 2-11. Relative frequencies of play behavior by age and sex classes.

Behavior		Adolescent Male	Adolescent Female	Juvenile Male
(BO)	Bite other	2.53	0.13	1.97
(CH)	Chase	9.46	0.46	0.81
(ES)	Escape	0.00	1.53	0.00
(FR)	Forward roll	0.00	1.40	0.00
(GR)	Grab other	8.46	1.33	3.77
(PL)	Pull	1.33	0.13	0.55
(RL)	Roll	0.13	1.26	0.00
(TC)	Touch	1.46	0.20	1.12

Although there were no sex or age differences in the mode and intensity of play for young orang-utans, the sex and developmental differences in the component behaviors link these play behaviors to the sexual behaviors of adult orang-utans. The typical pattern of male behaviors during copulation consists of chasing, grabbing, and restraining the female, including the biting of her extremities. All of these permit eventual vaginal penetration, and ejaculation. The female, on the other hand, initailly resists the male's advances, and once restrained by the male, continues to struggle throughout the length of the copulation bout. The sex differences observed during play bouts correspond directly to the adult copulatory behaviors of this species. Males chase, grab, and bite at a much higher rate than do the females, as well as pull the females from a hanging position more frequently than the reverse. In addition to these differences, older males (at or near maturity) emitted these behaviors at a relatively higher rate than did juvenile males. Escape behaviors, those that functioned to break contact with the other animal, were emitted exclusively by females. This, too, corresponds directly to the adult female's behavior during copulation. For all of these behaviors, the individual variability is great, but there is very little overlap in the range of these rates across age and/or sex classes. No *overt* sexual behaviors, such as thrusting or genital examination/contact, occurred during the heterosexual play bouts.

When these differences are considered with respect to the adult copulation sequence, the label "rape" is not totally accurate, even in apparently forceful copulations. If the copulating adults have had any heterosexual peer experience in the course of their development, which is presumed to be the case in the wild (cf. MacKinnon, 1974; Horr, 1978), then these animals have utilized in play the same behaviors that occur during copulation bouts. That the female struggles to free herself from the male's grasp cannot be conclusively interpreted as an attempt to thwart the male's sexual advances, but such behavior may well be gender-typical for this species when first encountering a male conspecific. The young adult virgin female encountering her first approaching mature male presumably has a knowledge of what behaviors will ensue due to her past play experiences. Whether or not she expects a play bout to occur cannot, of course, be determined. There are undoubtedly other communicatory signals such as facial expressions and vocalizations used in conveying the intent of the encounter. The putative "play face" is one such signal, although for orang-utans, it is an ambiguous expression due to its transitory nature—it is intermediate between any other expression and a bite. Bites, however, are a component of orang-utan play for both sexes, though they were emitted more frequently by males. Similarly, *hand extension* occurs as a communicatory gesture in both contexts (cf. Zucker *et al.*, 1978; Maple and Zucker, 1978).

To say that play functions solely to allow for the practice of the component behaviors of copulation is misleading. Play is undoubtedly a multifunctional constellation of behaviors (cf. Smith, 1978), and the context is an important determinant of the particular function at that particular time. However, our data support one of the hypotheses regarding the function of play, as we have seen. The developmental differences we observed indicate that the component behaviors of copulation are present in the repertoires of young orang-utans but the rates of these behaviors change as the animals mature. That is, new behaviors do not emerge at the time of sexual maturity, but the tendency to emit these behaviors changes over time. The point in ontogeny at which these changes occur, or begin to occur, is currently unknown. Continued longitudinal, or cross-longitudinal, studies of these young animals will provide some answers to these questions, possibly providing for an age classification of animals based on behavioral and social abilities, which would complement a classification system based solely on physical attributes.

Conclusions

From our studies of a large number of orang-utans, we have learned that the components and positions assumed in captive play bouts are basically similar throughout the lifespan of the animal. Both males and females emit rough and tumble play, and adults play frequently with each other and their offspring. Although not exhaustive, our data have allowed us to generate an ethogram of play for the captive orang-utan and to compare it to the behavioral repertoire of other great apes. In many ways, orang-utan play resembles gorilla and chimpanzee play, but there appear to be distinct differences as well. Many of these may be explained in terms of differences in bodily and social structural adaptations. Since we are attempting to study the great apes under similar conditions, we believe that our research will eventually yield valid comparisons of social behavior and social development. This effort can only be properly understood in relationship to the data acquired from field studies of each respective species. Play behavior, in particular, is likely to be overly represented in the repertoire of captive animals, but we trust that its form and function is similar in captivity and in the wild. Our studies are no substitute for data from the wild, and we aim to render our data comparable to field data.

3
Expression and Emotion in Orang-utan

> The faces of the more intelligent orangs are capable of a great variety of expression and in some the exhibition of the various passions which are popularly supposed to belong to human beings alone is truly remarkable.
>
> (Hornaday, 1879; p. 442)

There appear to be great differences in expressive behavior when the great apes are compared. As the reader will see, however, there are few objective data with which to describe these differences. Nonetheless, the experienced observer of apes cannot deny them. It is clear that the chimpanzee is the more expressive of the family. By "expressive" I mean here that this ape exhibits a greater *responsiveness* to external stimulation, and that the variety and intensity of its response repertoire exceeds that of the other apes. To be sure, there are occasions in which gorillas and orang-utans are as responsive as chimpanzees, but it generally takes more stimulation to arouse them to action. In this chapter, we will review some early impressions of orang-utan expression and emotion, the expressive repertoires of orangs as described by field workers, and some data acquired from my own research on captive animals. The reader should recognize that there is still much to learn about this aspect of orang-utan behavior. However, as will be seen, there is considerable agreement among observers regarding the basic elements of orang expressivity.

FACIAL EXPRESSIONS

Of all the great apes, orang-utans are capable of the greater degree of facial expressiveness. However, the plasticity of the orang face is what permits these various contortions, and it is not likely that they possess a greater number of discrete social signals. It is possible that the plastic face is an adaptation to arboreality whereby the "prehensile" lips aid in feeding while suspended. In view of the locomotor caution of the orang, such an adaptation would be extremely important. As Davenport (1967) discovered, orang-utans invariably maintain contact with their environment by grasping with three limbs.

> Regardless of the method of locomotion however, the animals always appeared to be careful, that is, at any one time three of the four extremities were grasping or in firm contact with a support. (p. 251)

About the lips, Robert and Ada Yerkes have written the following:

> The lips are conspicuously expressive because of their extreme mobility and constant use. They seem to play a far more important role in this respect in orang-utan and chimpanzee than in gibbon, gorilla, or man. From our own observations we should conclude that in contentment or delight they are protruded slightly if at all, and they may be held firmly together or parted in less or greater degree; but in impatience or resentment, in begging or mute appeal, they commonly are protruded in funnel shape as if to facilitate the peculiar vocalization characteristic of these attitudes. As resentment gives way to anger or rage they may be protruded extremely in accompaniment with loud screaming, or the mouth may be opened widely and the lips drawn back. Undoubtedly in all such emotional conditions marked play of feature occurs; yet no description of it is available. Truly there is much to see, much to measure, much to describe, correlate and explain in orang-utan affective behavior whether or not feeling, emotion, mood, or sentiment be present. (1929 p. 160).

Nevertheless, prehensile face notwithstanding, orang-utans do emit expressions which are distinctly communicative. Historically, naturalists recognized this expressiveness from their experiences with young animals. Darwin, however, seemed to be somewhat confused about the degree to which both orangs and chimpanzees were expressive. Consider the following two statements extracted from his 1872 volume *Expression of the Emotions in Man and Animals*:

> The appearance of dejection in young orangs and chimpanzees, when out of health, is as plain and almost as pathetic as in the case of our children. This state of mind and body is shown by their listless movements, fallen countenances, dull eyes, and changed complexion. (p. 134)

> Although the countenances, and more especially the gestures, of orangs and chimpanzees are in some respects highly expressive, I doubt whether on the whole they are so expressive as those of some other kinds of monkeys. This may be attributed in part to their ears being immovable, and in part to the nakedness of their eyebrows,

of which the movements are thus rendered less conspicuous. When, however, they raise their eyebrows their foreheads become, as with us, transversely wrinkled. In comparison with man, their faces are inexpressive, chiefly owing to their not frowning under any emotion of the mind—that is, as far as I have been able to observe, and I carefully attended to this point. (p. 141)

Of course, the quantitative comparison of primate taxa on some dimension of expressiveness is exceedingly difficult. Perhaps it is best to view *expression* in the same way that we now view *intelligence*. In doing so, we acknowledge that *expression*, like *intelligence*, is precisely adapted to the ecological niche and survival problems of each respective species. There are quantitative and qualitative differences among the primates in expression, but these are best understood within the framework of adaptation. Once complete descriptions have been acquired, direct comparisons can be made. We will concentrate first on the former, although some historical statements, such as Darwin's, are inherently comparative as phrased.

As MacKinnon (1974) has pointed out, forest conditions are not ideal for the study of orang-utan facial expressions. Because of these problems in visibility, MacKinnon's list of expressions and gestures was incomplete and lacked sufficient detail (cf. Table 3-1). He suggested that a complete ethogram could be accomplished by carefully

Table 3-1. Facial expressions of orang-utan as recognized by MacKinnon and Rijksen.

MacKinnon	Rijksen	
eye flash		when eyes blink closed, pale pink patches are most visible; possibly appeasement
fear face	horizontal bared-teeth face	grimace; gums exposed, upper lip pulled up, nose wrinkled (MacK.) retraction of mouth corners and lips exposing teeth and gums; submission (Rijk.)
bared-teeth threat	open mouth bared-teeth face	exaggerated yawn, showing teeth (MacK.) teeth exposed, mouth open; appeasing during play (Rijk.)

Table 3-1. Facial expressions of orang-utan as recognized by MacKinnon and Rijksen (continued).

MacKinnon	Rijksen	
	silent-pout face	lips pushed forward while pressed together at mouth corners to form a small aperture; may be submissive request for tolerance or appeasement
	†tense-mouth face	lips closed tightly during tense social interactions; similar in chimpanzee and gorilla
	bulging lips	lips pressed together but pulled inward with face bulging outward; emitted by females during copulation
	frowning	only observed at close quarters during attack; aggressive expression
	fixed gaze (stare)	alternates with looking away at right angles as in gorilla; threat gesture
play face	relaxed open-mouth face	mouth open, lips curled back tight (MacK.) widely opened mouth, withdrawn corners, teeth often covered by lips; observed during play (Rijk.)
pout face	pout moan face	mild fear, accompanies mild whimper, lips like trumpet (MacK.) lips pouted in funnel; distress face (Rijk.)
*kiss face	*kiss face	extremely pouted lips in funnel shape; agonistic display of threat and fear, perhaps diverts attention from young
inflation of laryngeal pouch		body appears larger; intimidation

*Depicted but not actually discussed by MacKinnon and Rijksen as a facial expression.

†Rijksen found that the *bulging lips* face of Sumatran orang-utans occurred in females during copulations. Since the chimpanzee equivalent occurs in aggressive situations, it would seem that the orang equivalent is not simply an "exaggerated tense face" as in chimps. However, given the tension that generally accompanies rough-and-tumble orang copulations, I am inclined to see the orang and chimpanzee "bulging lips" face as essentially equivalent and hence homologous expressions.

studying rehabilitants or zoo animals. Thus, support for zoo studies is again provided. We cannot rely on field work to supply all of the answers to our questions. Some problems are best studied under captive conditions where repeated observation under controlled conditions is possible.

EXPRESSIONS OF WELL-BEING

In orangs, chimpanzees, and gorillas bodily stimulation such as *tickling* or rough and tumble play will produce expressions and vocalizations which resemble human laughter. Modern scientists, such as Marler (1967) and Goodall (1968) have readily applied this term, as did Darwin.

> . . . Young orangs, when tickled, likewise grin and make a chuckling sound; and Mr. Martin says that their eyes grow brighter. As soon as their laughter ceases, an expression may be detected passing over their faces, which . . . may be called a smile. (1872, p. 132)

The Yerkes' were somewhat skeptical about our state of knowledge concerning orang-utan expressiveness. They wrote as follows about *joy*:

> A characteristic behavioral picture of joy has not been drawn. No one it seems has observed it with sufficient patience, skill, and objectivity to describe even the facial changes, still less the bodily attitude and physiological conditions. It is a task worth the attention of a first-rate observer. (1929, p. 159)

Skepticism aside, the Yerkes' did acknowledge the existence of a laugh-like expression in orangs.

> Often in play with their fellows or with persons, and when chased, mauled, or tickled, young orang-utans chuckle, grunt, gurgle, and exhibit distortion of facial features which might readily be taken, or mayhap mistaken, for smiling or laughter. (p. 159)

Of further interest in this issue is an early description of *laughter* offered by Grant in 1828 as follows:

> Although the beautiful play of the features which we call smiling is confined to man alone, yet is the orang-utan capable of a kind of laugh when pleasantly excited. For instance, if tickled, the corners of his mouth draw up into a grin; he shows his teeth,

and the diaphragm is thrown into action, and reiterated grunting sounds, somewhat analogous to laughter, are emitted by the animal. The creature indeed is extremely sensitive to tickling in those parts where a human being is, as the armpits and sides. (p. 4)

In the recent literature, MacKinnon (1974) described only the vocal component of laughter. However, he did describe the orang *play-face*, commonly attributed to apes.

In appearance the play-face is similar to the fear-face but it is shown in quite a different context. It occurs during intense play between young animals and even immediately prior to joining in play where it apparently functions as a play invitation. It was occasionally also seen during auto-play. The mouth is open and the lips curled back tight . . . the teeth may be covered but usually at least one set is exposed and used in play-biting. The nose is screwed up close to the eyes, exaggerating the facial creases. (p. 61)

Elsewhere (Maple and Zucker, 1978) we have also discussed the play-face, arguing that its communicative significance is questionable. That it occurs during play in apes, however, cannot be denied.

In his study of free-ranging Sumatran orangs, Rijksen described those facial expressions which occurred during *gnaw wrestling* bouts.

During 'gnaw wrestling' both participants regularly showed a 'relaxed open-mouth' facial expression similar to that described for the other great apes. (p. 215)

Rijksen also described an *open-mouth bared-teeth face* where the corners of the mouth were drawn back, exposing the teeth and sometimes part of the upper jaw with mouth wide open. The expression was emitted often by young animals, especially rehabilitants, but Rijksen saw it only once in an adult.

Typically lower ranking individuals reacted to the final phase of a 'play approach' of a higher ranking orang-utan, with this 'open-mouth bared-teeth' facial expression just prior to making contact, apparently as an acceptance of the play invitation. . . . Its context suggests that this expression bears an appeasement message by neutralizing elements that could look assertive (such as the higher spatial position) or aggressive (such as the 'arm wave' and 'grasp'). (p. 221)

Rijksen went on to discuss some distinctions between *open-mouth bared-teeth, horizontal bared-teeth,* and *relaxed open-mouth* facial expressions. However, because he presents no quantitative evidence for these distinctions, there is little reason to accept them at this time. It is sufficient to say that "playful" interactions often involve components which suggest fear, fight, and fun. To distinguish between them requires, at the least, a careful motion-picture analysis of the component motor-patterns, expressions, and vocalizations.

EXPRESSIONS OF PAIN AND FEAR

I wish to emphasize again here that the breakdown of expressions into discrete categories and elements is difficult. We hope that it isn't also arbitrary. The scientist who studies expression attempts to derive categories which are biologically meaningful units, that is, units which are meaningful to the animals we study. The process of analysis and component breakdown is tedious. As in the case with the previous category, good qualitative descriptions exist, but a quantitative motion picture analysis has yet to be published.

Charles Darwin (1872) described an expression for orang-utans which he labeled as *surprise*. He compared this expression to that emitted by humans under similar circumstances.

> . . . although when thus affected, our mouths are generally opened, yet the lips are often a little protruded. This fact reminds us of the same movement, though in a much more strongly marked degree, in the chimpanzee and orang when astonished. As a strong expiration naturally follows the deep inspiration which accompanies the first sense of startled surprise, and as the lips are often protruded, the various sounds which are then commonly uttered can apparently be accounted for. (pp. 284–285)

Surprise easily gives way to fear, in fact, Darwin's description of the *protruded lips* expression contains as much *fear* as *surprise*. Being mindful of the many different labels, another example of the *protruded lips* face is called *violent alarm* by Abel.

> On seeing eight large turtle brought on board . . . he climbed with all possible speed to a higher part of the ship than he had ever before reached; and looking down upon them, projected his long lips into the form of a hog's snout, uttering at the same time a sound which might be described as between the croaking of a frog and the grunting of a pig. After some time he ventured to descend, but with great caution, peeping continually at the turtle, but could not be induced to approach within many yards of them. He ran to the same height and uttered the same sounds on seeing some men bathing and splashing in the sea; and since his arrival in England, has shown nearly the same degree of fear at the sight of a live tortoise. (1818, p. 329)

The *protruded lips* display, because of the orang's facial plasticity, has often been described as an intention to *kiss*, as in the following statement by Darwin in 1872.

Many years ago, in the Zoological Gardens, I placed a looking-glass on the floor before two young orangs, who, as far as it was known, had never before seen one. At first they gazed at their own images with the most steady surprise, and often changed their point of view. They then approached close and protruded their lips toward the image, as if to kiss it, in exactly the same manner as they had previously done towards each other, when first placed, a few days before, in the same room. They next made all sorts of grimaces. (p. 140)

This display may indeed contain elements of fear, but it seems also to be emitted as a threat. We will look at it in this context in the section to follow. MacKinnon (1974) reported that protruded lips were, indeed, an indication of fear. MacKinnon's name for this expression was the *pout-face* which he described thus:

The pout face is shown in contexts apparently indicating mild fear. It is often accompanied by soft whimpering in young animals or 'kiss squeaks' in adults. The lips are pursed forwards and parted at the end to form a trumpet. (p. 61)

Similarly, Rijksen (1978) suggested that this expression signified a "submissive request for tolerance or appeasement." (p. 223)

Of course there are other expressions which are indicative of pain and fear. About the former, which is fraught with anthropomorphic overtones, it is not possible to improve upon this 1929 remark by the Yerkes'.

Although indications of depression, sadness, and grief are not wanting, here also useful descriptions are lacking. Since naturally the orang-outan tends to quiescence and serious mien, it is inevitably characterized as melancholy in appearance, and indeed, except for moments in infancy and childhood when the joy of living seems to possess it, it behaves as though depressed. There are, however, depths and degrees of depression, and it is possible by various means to induce even more marked expression of sadness or grief than the ordinary. In disappointment the young specimen quite commonly whimpers or weeps, without, however, shedding tears. Only casual references to such affective behavior have been found and they contribute little and uncertainly to our knowledge. (p. 161)

MacKinnon (1974) describes a *fear-face* as a grimace in which the sides of the mouth are pulled back, gums exposed, and the upper lip

Figure 3-1. Orang-utan expressions.
(a) Juvenile's *horizontal bared-teeth face*;
(b) Adult female's *silent bared-teeth threat*;
(c) Adult female's *kiss-squeak* vocalization;
(d) Infant's *open-mouth bared-teeth face*;

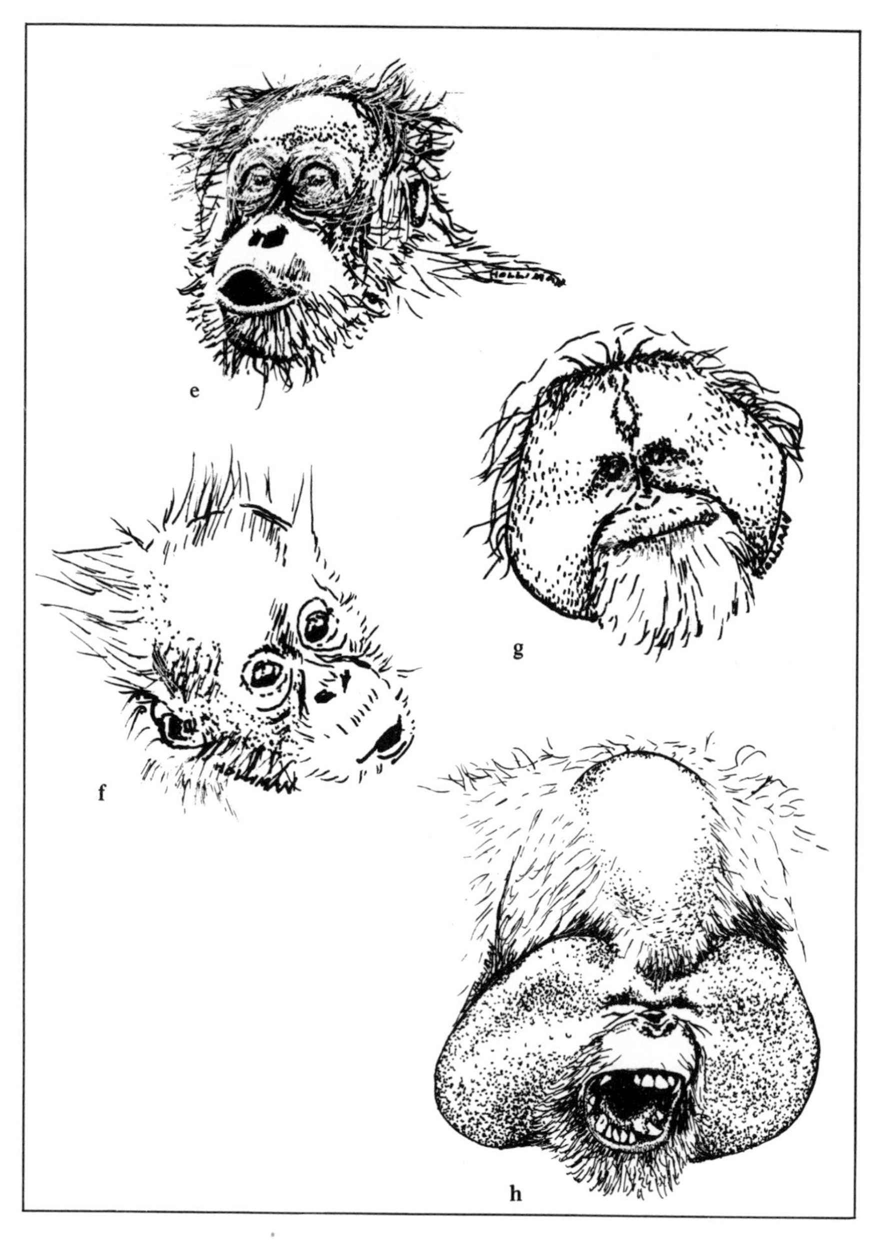

(e) Juvenile's *silent pout face*;
(f) Infant's *pout-face* when whimpering
(g) Adult male's *tight-lipped tense face*;
(h) Adult male's *wide open mouth bared-teeth face*.

All figures drawn by B. Holliman from photographs by MacKinnon (b, c, f) and Rijksen (a, d, e, g, h).

Figure 3-2. A hypothetical hierarchy of fear expressions in the orang-utan.

pulled up to crinkle the skin around the nose. Since screams and defecation often accompanied this face, MacKinnon's category does not lack validity.

An expression of tension, with lips tightened, is common to humans, gorillas, chimpanzees, and orang-utans. However, in the more stoic orang, the expression is seen less frequently. As Darwin noted:

> When we try to perform some little action which is difficult and requires precision, for instance, to thread a needle, we generally close our lips firmly, for the sake, I presume, of breathing; and I noticed the same action in a young Orang. The poor little creature was sick, and was amusing itself by trying to kill the flies on the window-panes with its knuckles. This was difficult as the flies buzzed about, and at each attempt the lips were firmly compressed, and at the same time slightly protruded. (1872, pp. 140–141)

The retraction of the upper lip, and also a chomping of the teeth with lower jaw thrust out, are subtle movements of the mouth which we have correlated with events of minor irritation. I would characterize these expressions as transient indicators of stress. The latter may be a mild form of threat, as it is often accompanied by glass and bar-pounding.

VOCALIZATION

As all observers of orang utans have noted, they are relatively quiet creatures. MacKinnon (1971) noted that ". . . their vocalizations are only audible at close quarters and several are made by inhaling rather than exhaling."

A recent study by Niemitz and Kok (1976) examined the vocal behavior of a confiscated infant orang-utan residing at the Sarawak Museum. This report is especially notable in that the ontogeny of orang-utan vocal behavior is a virtual mystery. The investigators recorded vocal behavior and analyzed the recording by the use of a sonograph. The subject emitted its characteristic vocalization after being "frustrated" by the investigators. To prevent any psychological harm, the experiment was carried out only twice. The investigators speculated about the communicative significance of the infant's frustrated cries but could not verify their suggestion that mothers might respond to such utterances. However, it is clear to me, from my own observations, that mothers retrieve and attend to squeaking, and screaming infants. Moreover, we have found that persistent would-be foster mothers often cease retrieving a newly adopted infant if it emits loud vocalizations. Thus, these cries may be seen as "contact" calls (cf. Mitchell, 1979), but may also deter rough or novel (hence frightening) contact. Hand-raised orangs typically emit a persistent squeak at the sight of their caretaker, clearly seeking ventral contact. Moreover, when they have been abandoned by their caretakers their squeaks quickly become loud screams, punctuated by terminal grunts. Genuine *tantrums* are common for baby orang-utans.

TEMPERAMENT

It is difficult to characterize the emotional makeup of any animal. Under appropriate circumstances, the most placid animal can be highly aggressive, and the most taciturn animal can be congenial in its relations with others. When we speak of the orang-utan's "personality" or temperament, we can rely on centuries of generalizations or one recent objective test.* Let us begin by reviewing the first source of information.

It is important to note at the outset that there are vast differences in the temperaments of young and old orang-utans. Infants and juveniles are generally active and responsive, whereas adult animals are often exceedingly lethargic to the point of sitting motionless for hours on end. To a degree, this characteristic adult lethargy is due to the stupefying habitats in which captive animals must often live (cf. Chapter 7), but wild adults are also somewhat less active than their younger counterparts. From observation of captive specimens, the Yerkes' (1929) also made distinctions by age.

The temperament of the orang-utan has been described by many naturalists and scientists, both from captive and field encounters (see also Chapter 6). For example, Rennie (1838) refers to the observations of a Dr. Muller who frequently observed them in Borneo:

* My more clinically oriented friends may be surprised to learn that I am at work on the development of a psychological test which I call Maple's Multiphasic Pongo Personality Inventory (MMPPI).

Table 3-2. Comparison of vocal labels as suggested by Rijksen and MacKinnon.

Rijksen (1978)	MacKinnon (1974)	Composite function/context
squeak	fear squeak	appeasement gesture
ahh	play grunt	"laughter" (Rijksen); accompanied by "play face"
grunt		low intensity, may be a feed call; significance unclear (Rijksen)
contact uff		"restrained squeak vocalization" (Rijksen)
grumbling		soft, dull sound; agonistic; made by both sexes (Rijksen)
pout moan	soft hoot/whimper	distress call (Rijksen); made by frightened infants (MacKinnon)
bark	bark	possibly a startle call (Rijksen); possibly threat or warning (MacKinnon); uncommon
mating squeaks	mating cries	long, drawn out squeaks by females during copulation (Rijksen); rhythmical fear screams (MacKinnon)
scream	crying and screaming/ frustration screams	uttered by young animals only for attracting attention, persuasion (Rijksen); MacKinnon differentiates between screams at separation and screams at frustration
bared-teeth scream		uttered during rape or when attacked
grumpf	grumph	often accompanies kiss; often a tendency to flee, occurs when annoyed (Rijksen); belch-like grunt, second commonest calls heard (MacKinnon)
	gorkum	transition from grumph to lork; intimidation display (MacKinnon)
Lork	Lork call	subadult male intimidation call, adult female advertisement (Rijksen); violent intimidation call by annoyed animal, made mainly by females (MacKinnon)

Table 3-2. Comparison of vocal labels as suggested by Rijksen and MacKinnon (continued).

Rijksen	MacKinnon	Composite function/context
long call	long call	loud intimidation call by adult males (Rijksen); functions to advertise territory, warn other males, attract females (MacKinnon)
chomping	chomp	chewing movement, using tongue to produce sound (Rijksen); rhythmic gulping sounds with closed mouth (MacKinnon)
spluttering	raspberry	blowing air through compressed lips (Rijksen and MacKinnon)
grinding		grinding of teeth when frightened, emitted by rehabilitants and captives
kiss sound	kiss squeak/ wrist kiss	sucking in air through pouted lips, uttered when annoyed by mildly aroused individuals (Rijksen); made toward humans and when chased by another orang (MacKinnon)
	Ahoor call	adult male intimidation display (MacKinnon); uncommon call

He describes them as being in the highest degree unsociable, leading, for the most part, a perfectly solitary life, and never more than two or three being found in company. Their deportment is grave and melancholy, their disposition apathetic, their motions slow and heavy, and their habits so sluggish and lazy, that it is only the cravings of appetite, or the approach of imminent danger, that can rouse them from their habitual lethargy, or force them to active exertion. When under the influence of these powerful motives, however, they exhibit a determination of character, and display a degree of force and activity, which would scarcely be anticipated from their heavy, apathetic appearance. (p. 120)

While Robert and Ada Yerkes (1929) summarized the "total affect" of the orang-utan with the phrases ". . . lack of ambition, of aggressiveness, of determination, and of energy; discouragement, pessimism . . ." they also reflected on the difficulty of generalizing from one individual to another.

. . . it is certain that no single, simple, or even complex formula for temperament is equally applicable to all individuals. It is definitely known that in both amount and character, activity varies extremely with age and stage of development, the young being relatively active, energetic, playful, and the old, sluggish and content to rest in quiet. In addition, scarcely less marked are the differences characteristic of individuality, sex, particular modes of experience, and pathological conditions. (p. 151)

To further amplify their views, the Yerkes' commented on the then current state of knowledge of orang-utans as follows:

As one observes for himself, and seeks to supplement his firsthand information by searching the literature, he comes to suspect that adequate general description of the affective behavior of the orang-utan is a difficult and intricate task which up to the present has not been seriously undertaken by any adequately trained psychologist. (p. 151)

Sex differences in temperament, although inferred from a small sample, were also suggested by the Yerkes'.

Our own acquaintance with the orang-utan is limited to four captive specimens. The first, a male of approximately five years, used especially as subject for experimental studies in adaptation of behavior, was eminently childlike in his expressions of attachment to the observer, and has been characterized as gentle, docile, friendly . . . By contrast with two females, probably between six and eight years of age and obviously nearing sexual maturity, the five-year-old male was markedly more active, energetic, and playful (see p. 154).

The presumed sex differences in playfulness have also been detected in our recent research (Zucker *et al.*, 1977) in which young males have been found more likely to be initiators of play, and more likely to grasp and bite during play whereas females tend to resist and flee from the males, and emit fewer biting/grasping behaviors during such encounters. While it is wrong to suggest that females are less playful, their *mode* of play appears to be different. We have argued that those behaviors which differentiate the sexes in youthful play correspond to later sex-specific behaviors during heterosexual bouts of copulation, as described in Chapter 2.

In a comparative perspective, we have attempted to determine the differences between chimpanzees, gorillas, and orang-utans in their general level of temperament. It is important to note here that there have been few instances where experienced observers have *objectively* compared the apes in this dimension. For example, Sonntag (1924) wrote:

> The orang is the least interesting of the Apes. It lacks the grace and agility of the gibbon, the intelligence of the chimpanzee, and brutality of the gorilla. (p. 80)

As the Yerkes rightly commented ". . . this form of statement is out of place in a scientific treatise" (p. 111). Nonetheless, even the Yerkes' did not have sufficient direct experience with all of the anthropoid taxa, and were unable to provide data-based support for their comparative generalizations. Undaunted, however, they argued as follows:

> No assemblage of psychobiological characteristics is more difficult to describe briefly and safely than the affective. Language proves inadequate for description of conditions and indication of contrasts. Nevertheless, attempt is made . . . to characterize briefly the affective life of each type and to make comparisons. (p. 544)

In so doing, the Yerkes divided their section "affective traits" into six basic categories: (1) emotionality and expressivity; (2) dispositional contrasts; (3) variety of emotional pattern; (4) vocalization and sound production; (5) speech and other means of intercommunication; (6) motivation. For the first category, the Yerkes argued that if expressivity is used to determine order of *increasing* emotionality the order for great apes would be: gorilla, orang-utan, and chimpanzee. For category two, the three taxa were compared as follows:

> The orang-utan, quiet, inactive, or sluggish, gives the impression of stolidity, brooding, depression, melancholy. It is phlegmatic and its attitude and behavior often strikingly suggest pensiveness and pessimism. Strong indeed is the contrast between this picture and that of the chimpanzee, for it is active, lively, sanguine, very highly expressive, with indications of nervous instability, restlessness, impulsiveness. The gorilla, although differing markedly from both, seems to be more like the orang-outan than the chimpanzee temperamentally. It is shy, retiring, discreet, deliberate rather than stolid, obviously self-dependent and self-centered . . . The order of increasing moodiness is . . . chimpanzee, orang-outan, gorilla. (pp. 544–545)

In category three, the Yerkes' looked at the expression of "shyness," characterizing the ascending order as gorilla, orang-utan, and chimpanzee. On an aggressiveness scale (composed of "resentment, anger, and rage") the increasing order was alleged to be gorilla, orang-utan, chimpanzee. So-called "agreeable" emotions, such as pleasure, contentment, satisfaction (manifest in smiling, laughter, humor, mischievousness) were said to progress in frequency from gorilla to orang-utan to chimpanzee. Finally:

Evidences of sorrow, grief, depression, sympathy, attachment, mutual and altruistic aid are . . . increasingly abundant in the order: orang-outan, gorilla, chimpanzee. (p. 545)

For *vocal* sound production, the Yerkes' determined that the increasing order for frequency was gorilla, orang-utan, and chimpanzee. But in order of frequency for *nonvocal* sounds, such as handclapping, the order was orang-utan, chimpanzee, gorilla.

Both hands and feet may on occasion be used by the chimpanzee to beat on the earth or on surrounding objects. That this is done in order to make a noise is highly probable. The gorilla, most taciturn perhaps of all the types, leads in nonvocal sound production, for it uses either clenched or open hands to beat upon its own body or on other objects and thus produces a considerable variety of sounds . . . While the chimpanzee appears to be a most gifted vocalist among anthropoid types, the gorilla evidently is correspondingly gifted in production of nonvocal sounds. (p. 545)

As the Yerkes stated, it cannot be asserted that any one of the anthropoid types speaks. However, in their fifth category, they argued that apes systematically used the voice as a means of expressing "feelings, desires, and ideas." The order of ascending propensity to do so they put as gorilla, orang-utan, and chimpanzee. Moreover, they considered nonvocal communication as follows:

Mutual understanding and transfer of experience among apes are dependent rather on vision than on hearing, for the animal reads the mind of its fellow, interprets attitude, and foresees action rather as does the human deafmute than as the normal person who listens and responds to linguistic vocalization. Inter-communicative complexity and biological value increase, we hazard, in the order . . . orang-outan, gorilla, chimpanzee. (p. 546)

Finally, concerning *motivation*, the increasing variety and complexity of motivational factors was suggested as orang-utan, chimpanzee, and gorilla.

I have constructed a summary of these six categories which can be found in Table 3-3. Clearly, a composite order of increasing "emotionality" would be orang-utan,* gorilla, chimpanzee. As common

* Will Cuppy (1939) also attempted to characterize orang-utan temperament as follows:
Both sexes brood a lot. Their prolonged spells of meditation appear to have no tangible results. Orangs often sleep on one arm and wake up with a cramp. They snore. Young Orangs who are permitted to develop their individualities turn out horribly. Young Orangs who are kicked and beaten into line also turn out horribly. (Will Cuppy, *How to Tell Your Friends from the Apes*, pp. 37–38.)

Table 3-3. Hypothetical composite of Yerkes and Yerkes 1929 great ape "emotionality" comparisons.

	1. Emotionality/ Expressivity	2. Dispositional Contrasts	3. Variety of Emotional Pattern	4. Vocalization and Sound Production		5. Speech/ Inter-communication		6. Motivation	Total Rank
				vocal	nonvocal	vocal	nonvocal		
Orang-utan	2	2	2	2	3	2	3	3	3 ($\sum = 19$)
Chimpanzee	1	3	1	1	2	1	1	2	1 ($\sum = 12$)
Gorilla	3	1	3	3	1	3	2	1	2 ($\sum = 17$)

sense would tell us, the chimpanzee is a runaway leader in manifest "emotions." But is there any way to quantify these impressions? My student Ron Schonwetter and I think so. To this end, we designed and carried out the following experiment.

Table 3-4. Subject Information.

Species	Name	Sex	Age	Rearing History
G. gorilla gorilla	Banga	F	14*	wild born, captive reared
	Jini	F	15*	wild born, captive reared
	Katoomba	F	16*	wild born, captive reared
	Inaki	F	11*	wild born, captive reared
	Oko	F	15*	wild born, captive reared
	Paki	F	15*	wild born, captive reared
	Calabar	M	15*	wild born, captive reared
	Ozoum	M	17*	wild born, captive reared
Pongo pygmaeus	Datu	F	18*	wild born, captive reared
	Guchi	F	12	captive born and reared
	Ini	F	19*	wild born, captive reared
	Lada	F	20*	wild born, captive reared
	Paddi	F	19*	wild born, captive reared
	Tupa	F	20*	wild born, captive reared
	Bunga	F	7	captive born and reared
	Biji	F	8	captive born and reared
	Bagan	M	20*	wild born, captive reared
	Dinding	M	20*	wild born, captive reared
	Durian	M	29*	wild born, captive reared
	Lipis	M	21*	wild born, captive reared
	Sampit	M	22*	wild born, captive reared
	Dyak	M	20*	wild born, captive reared
	Loklok	M	9	captive born and reared
	Lunak	M	7	captive born and reared
P. troglodytes	Cheri	F	31*	wild born and reared
	Jenda	F	20	captive born and reared
	Maria	F	29*	wild born and reared
	Mary	F	18	captive born and reared
	Netta	F	adult	wild born, captive reared
	Maxine	F	21*	wild born, captive reared
	Wenka	F	24	captive born and reared
	Decamethonium	M	adult	captive born and reared
	Frans	M	32	captive born and reared
	James	M	adult	unknown
	Hoboh	M	adult	unknown
	Jimoh	M	adult	unknown
	Walnut	M	13*	wild born, captive reared

*estimated age

In our comparative study of *Gorilla gorilla, Pan troglodytes* and *Pongo pygmaeus*, numbers of each species were observed in order to distinguish similarities and differences in *temperament*. Behavior at feeding was used as an index to assess the range of emotional behaviors. Food is a basic requirement for survival and animals respond emotionally to the condition of hunger. Appetitive behaviors include body movements and changes in activity which lead to food acquisition and consumption (cf. Young, 1973). The purpose of our research was to describe the repertoire of emotional behaviors in these three species under identical conditions, in order to establish an objective index of *temperament* for each.

The subjects for this study were 37 great apes as described in Table 3-4. As stated previously, behavior prior to feeding was used as an index directly relating to the subject's *temperament*. The apes at the Yerkes Regional Primate Research Center respond to the sound of food carts by vocalizations and movement.

After a series of preliminary observations, specific behaviors were described for the three species. These behaviors, which included facial expressions, vocalizations and body movements, are listed in Table 3-5. A 10 key *Esterline Angus* Event Recorder was used to measure the duration and frequency of each behavior for one subject per cage as selected at random. Observations were recorded from an adjacent cage just ahead of the feeder for one-minute periods prior to feeding. The observer stood approximately 1 meter in front of the outdoor cages and each subject was observed five times.

From the tabulated data it was clear that the chimpanzee was by far the most expressive of the apes according to the measures we used. The orang-utan and the gorilla were fairly similar in the amount of expressive behaviors they exhibited. However, the gorilla expressed some behaviors which were absent in the orang-utan.

The common chimpanzee expressed all of the expected behaviors except *digit sucking*. Chimps were unique in exhibiting frequent loud *vocalizations, sexual excitement, hand-clapping, jumping* and *door/wall-pounding* just prior to feeding. The chimpanzee also exceeded the other two species in the duration of facial expressions observed, emitting distinct facial expressions almost 38% of the observation time. Erect penises were frequently observed in chimpanzees during the recording periods prior to feeding. Hand clapping also preceded

Table 3.5. Expressive behaviors recorded prior to feedings.

A. Funnel Face: Sometimes called a pout face, it consists of the full extension of pursed lips of the primate, with a parting at the ends to form a trumpet. This may accompany mild fear (MacKinnon, 1974).
B. Grimace or Open Mouth: Both facial expressions are quite similar and occur when either the corners of the mouth are pulled back, the mouth is slightly open forming a grin or when the teeth are showing but the lips are not pulled back. The skin around the nose seems to wrinkle. These usually accompany fear or play behavior, respectively (MacKinnon, 1974).
C. Loud Vocalization: Screaming, barking and usually long calls made by an animal.
D. Soft Vocalizations: Growling, squeaking, chomping, and gulping.
E. Sexually Excited: Observed only in males by penis erections.
F. Clapping Hands or Pounding Chest or Cheek: Making any kind of nonvocal sounds by slapping.
G. Digit-Suck: Animal sucking own finger(s) or toe(s).
H. Stationary Posture: Hanging, sitting, or stand (watching); three positions.
I. Aerial Movement: Brachiating or spider walking with either two arms or all four appendages.
J. Jumping or Wall/Door Pounding: Active bodily movements, or hitting, punching, or slapping door or walls with hands or legs.

feeding, and chimpanzees were the least stationary of the three apes, running, jumping, kicking and hitting the cages.

Gorillas, on the other hand, were more sedate and calm than the chimpanzees. Their facial expressions were the least frequent among the great apes studied. However, they too emitted some unique and distinguishable behaviors. They were unique by vocalizing in low-toned grunts or growls while waiting for food. This vocalization was observed over one-tenth of the test time. Species-typical chest beating was also observed, and gorillas exhibited the greatest amount of digit sucking. Overall, they were less active than the chimpanzees.

Interestingly, orang-utans exhibited the greatest interest in the recording device. They appeared to be as interested in this as with the food or feeder. They expressed slightly more facial expressions (about 2½%) than the gorillas. The only vocalization recorded was emitted by one adult male orang (long call). Some digit-sucking was observed, but not as much as in the gorillas. Orangs moved little prior to feeding.

Figure 3-3 contains the average percentage of time each *behavior* was observed in each species. Figure 3-4 contains the average percentage of time each *general category* was observed in the three species.

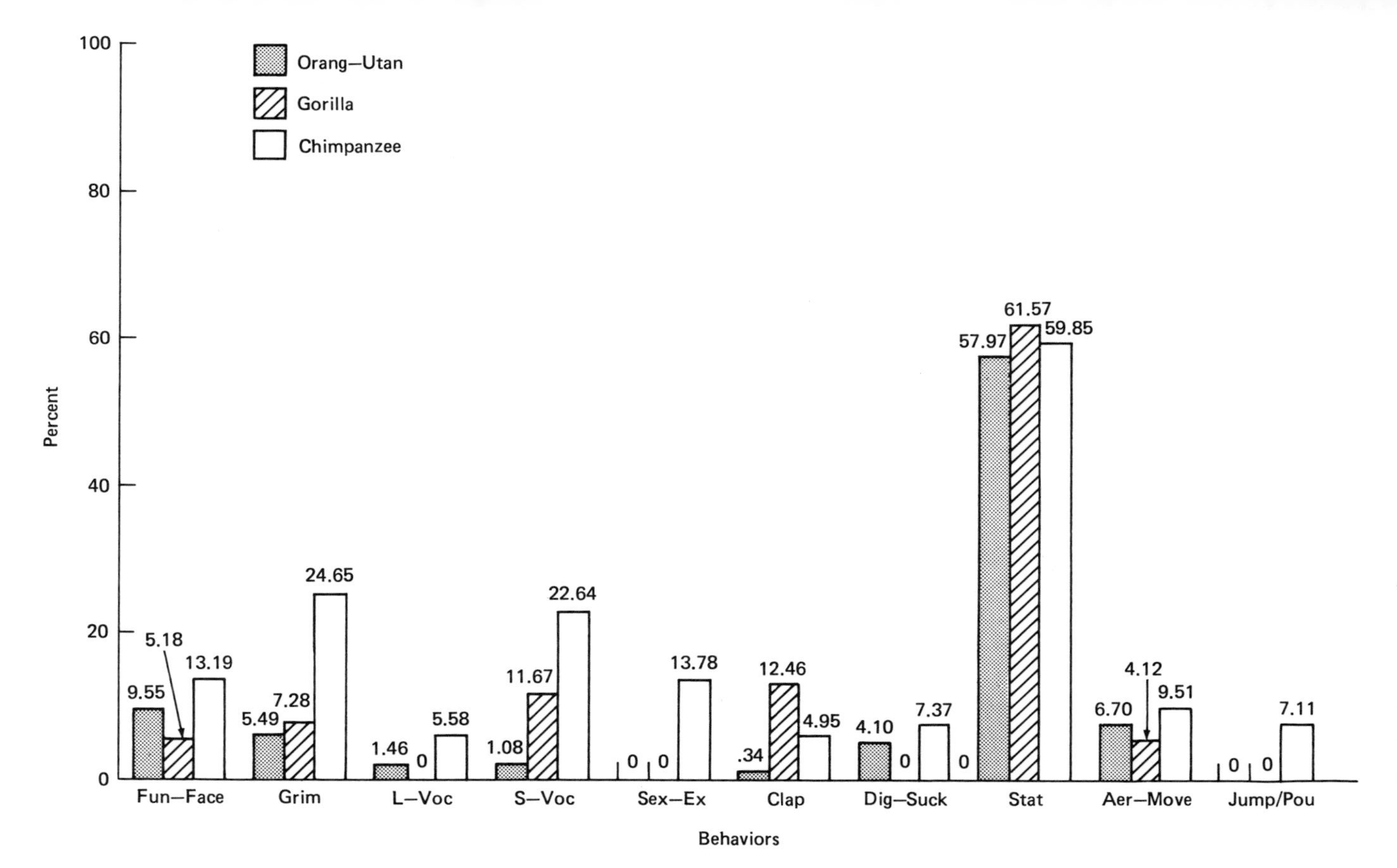

Figure 3-3. Percent of test that each behavior was observed.

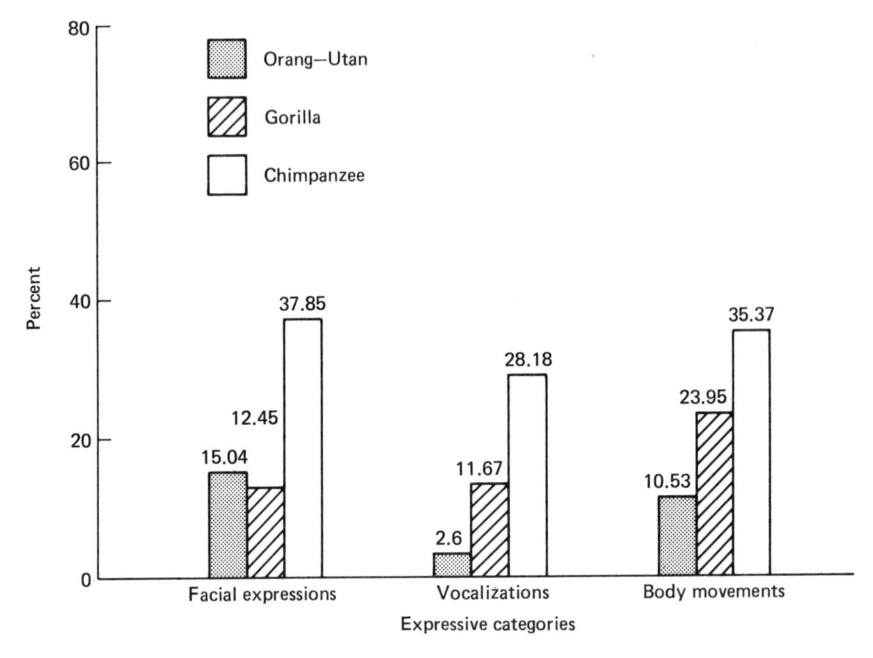

Figure 3-4. Percent of test that each general category of expressivity was observed.

Some unique patterns were noted when comparing behaviors of males and females of each species. In all three species the males scored higher in loud *vocalizations, brachiating, walking, jumping* and *door/wall pounding*.

In orang-utans and gorillas, females scored higher in all types of *facial expressions* as well as *soft vocalizations, hand-clapping, chest-pounding* and *digit-sucking*. Female chimpanzees, on the other hand, scored lower in *grimace, grin, open-mouth, soft vocalizations* as well as *hand-clapping, chest-pounding* and *digit-sucking*.

The juvenile orang-utans observed were more expressive than their adult counterparts. They scored higher than the adults in all behaviors except *grimace, grin* or *open mouth* and *loud vocalizations*. Overall they emitted more facial expressions, vocalized, and moved around their cages more than the adult orang-utans. Infant orang-utans observed during preliminary tests were all relatively inactive and quiet as they were still dependent on their mothers (not fed directly), and thus not conditioned to the cart and feeder. No infant or juvenile gorillas or chimpanzees were available for observation.

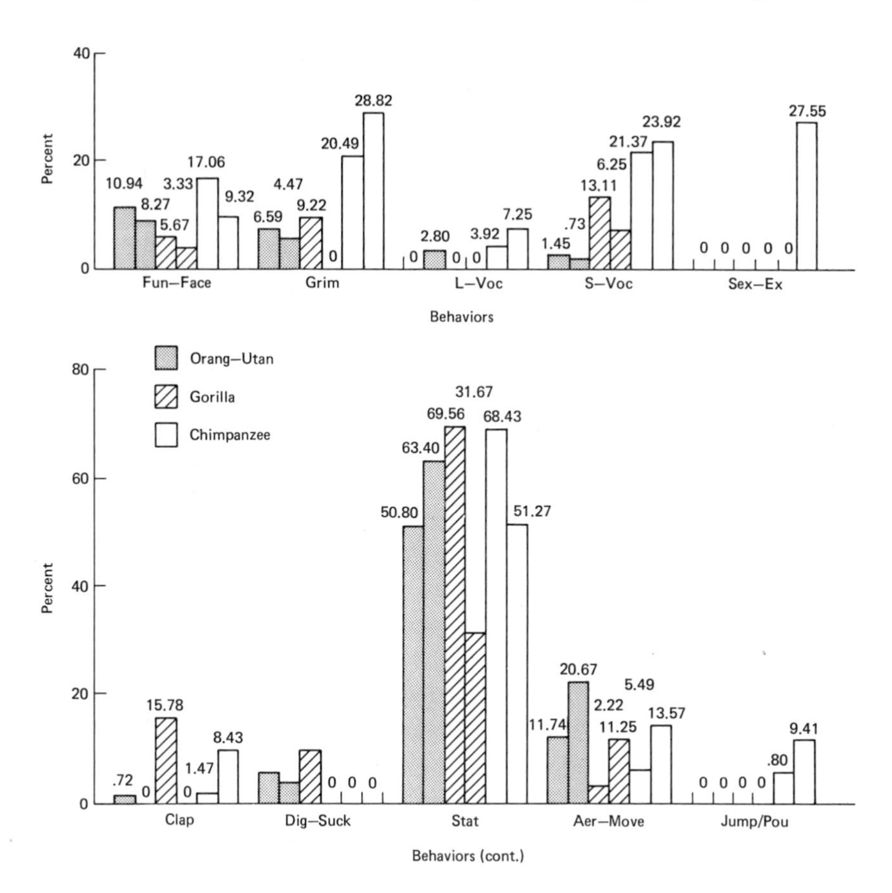

Figure 3-5. Differences in behavior due to sex.(First level is female; second level is male.)

Figure 3-5 compares the differences in behavior due to the sex of the three primate species and Figure 3-6 compares the behavioral differences due to the age of orang-utans only.

There were also some differences noted according to the social composition of the cages. When alone, orang-utans and gorillas seemed much less responsive as far as facial expressions, vocalizations and body movement were concerned. There was only one solitary chimpanzee studied, and it too was relatively quiet and inactive. When observed in peer groups of three, the orang-utans' behavior varied, but remained slow, while the gorilla groups were much more responsive than solitary or paired conspecifics.

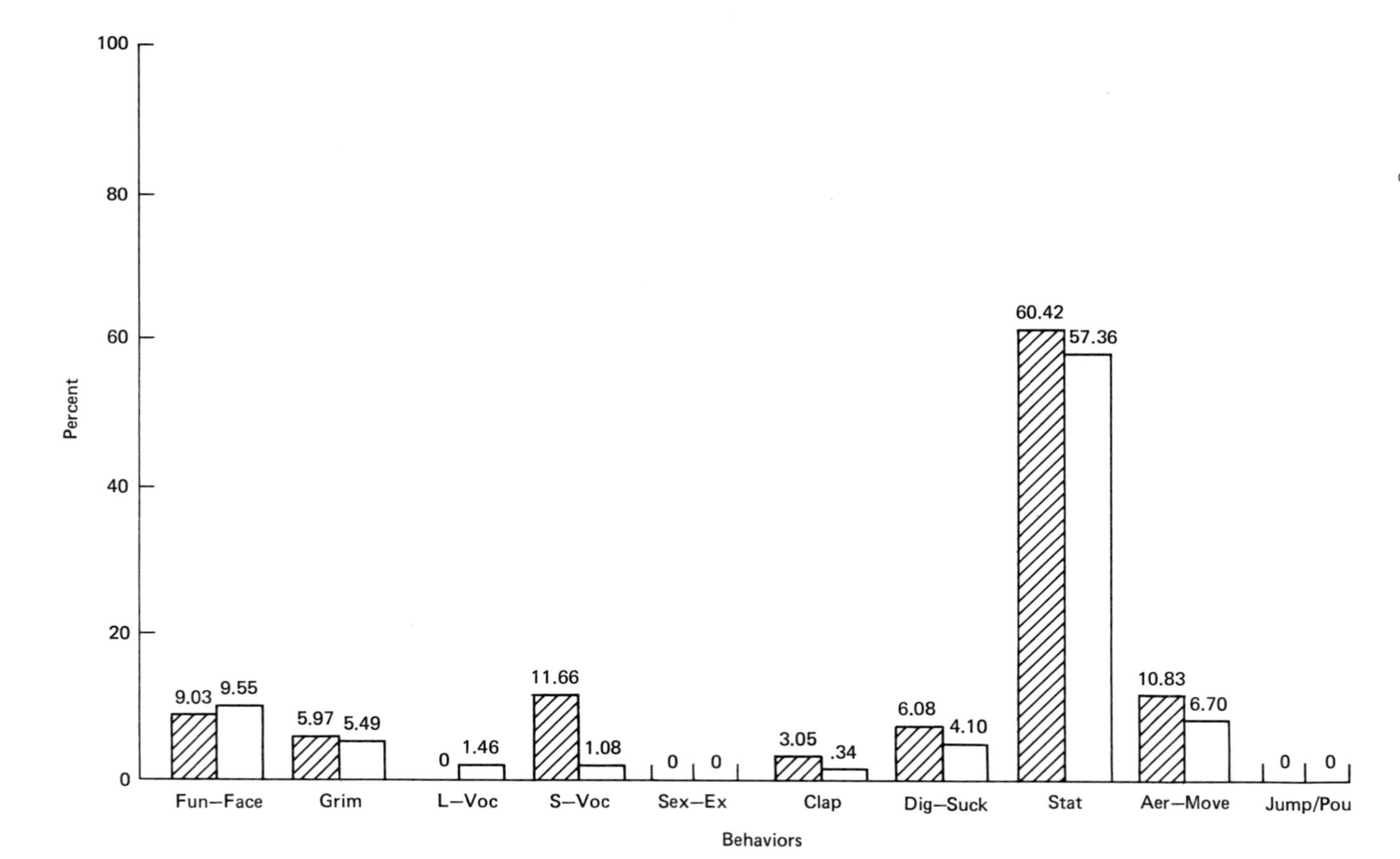

Figure 3-6. Differences in behavior due to age in orang-utans.(First level is juvenille, second is adult.)

Species-typical emotional behaviors are reliably emitted prior to feeding in chimpanzees, gorillas and orang-utans. If we search for an order of increasing emotionality among the great apes, the order obtained in this study was orang-utan, gorilla and chimpanzee. These results compares to the same composite order suggested by the Yerkes' (1929): orang-utan, gorilla, and chimpanzee. Our research reveals that orang-utans, gorillas and chimpanzees may be ordered in increased *variety* of expressiveness as well, a category in which the Yerkes' considered the orang to exceed the gorilla. The results of our study were very similar for orang-utans and gorillas. However, the gorillas exceeded orang-utans in *types* and *amounts* of bodily movements, and *types* and *amounts* of vocalizations, behaviors which were nearly nonexistent in the orang-utans. Facial expressions were seen slightly more in orang-utans.

As we previously suggested, age, sex, and social composition affected the behaviors recorded in this study. Better controls for social composition per cage, however, would have assisted in the determination of its effects on expressive behavior. Greater numbers of subjects would have also contributed to more valid comparisons in this study.

Whether or not social composition and its effect on expressivity had anything specifically to do with communication or competition between members of a cage could not be determined. Another study, using dependent variables other than responses to food, would determine whether the group increase in activity is a result of competition.

We also determined that the sex of the animal affects its expressivity at feeding. Males in the three species scored higher in many behavioral categories. Other behaviors such as *facial expressions, soft vocalizations, hand-clapping* and *digit-sucking* occurred to a greater extent in female orang-utans and gorillas. However, male chimpanzees exceeded females in most of these categories. The behavioral differences between male and female orang-utans and gorillas as compared to chimpanzees may be a result of differing dominance relations within the cages.

The expressions most common to each group have evolved and developed in response to the basic requirements of each species. The gorilla is shy, quiet and generally passive. It lives in stable groups in the wild and may not require an extroverted temperament. Its size and strength may also have led to decreased emotionality. The orang-utan is quiet, inactive and lethargic, and lives a relatively solitary life. The average group size is three members (cf. MacKinnon, 1974). Not quite as strong or as large as gorillas, orang-utans have developed other strategies for survival. Arboreal movement is quite common and a larger array of facial expressions for close communication have developed. Chimpanzees, on the other hand, travel in large, unstable groups and exhibit greater vocal communication. This may also be due to their relatively small size as compared to the other apes. Thus, their natural social organization and stature may well contribute to the development of particular species-typical behaviors.

The idea that food may mean different things to different species must also be considered. This brings forth the question of the validity of feeding behavior as an index of an animal's temperament. This study was designed to permit the animals to express themselves naturally in anticipation of food. However, under conditions of great deprivation, the orang-utan and gorilla might exhibit a similar degree of emotionality as the chimpanzee.

One may also argue that the appetitive behaviors observed were a result of operant conditioning in which the animals were rewarded with food for exhibiting unnatural behaviors. The flaw in this argument is that it is not important whether operant conditioning is involved, but that the behaviors were too similar within each taxon for one to say that they were unnatural. Thus, species-typical behavior appears to have been produced.

The recording device seemed to distract some of the primates, especially the orang-utans, and this may have affected the behaviors emitted. A less conspicuous means of recording or a longer adaptation period may be necessary in any further studies which are conducted.

In this study the order obtained for increasing expressivity and variety of emotional pattern was orang-utan, gorilla and chimpanzee. While previous researchers (.e.g, Yerkes and Yerkes, 1929) had based their temperament orders on subjective experience with members of the three species, the data we have reported are the first *objective* comparisons to be published.

4 Orang-utan Sexuality

> The orang-utan has always been painted as a creature of excessive libido and there are countless tales of night-long orgies in leafy nests, abducted Dyak maidens, passes made at zoo-keepers, and orangs kept for sexual entertainment.
>
> (John MacKinnon, 1974, p. 175)

Considerable interest has been recently focused on the sexual habits of the orang-utan. In this chapter I will describe the sexual behavior of orang-utans from data acquired in the laboratory, zoo, and field. The birth process, and emergent parental care will be described in Chapter 5 where the development of sexual behavior in immature animals will also be discussed. For all of these topics I will concentrate heavily on the studies which have been completed by my own research team.

Orang-utans, since their discovery by western observers over 300 years ago, have always been suspected of abducting human females. Although rumors of this proclivity are still credible among Indonesians (Galdikas, personal communication), the sexual aggressiveness of males is no exaggeration. In MacKinnon's field study of Bornean orang-utans (1974), seven out of eight observed episodes of sexual behavior were described as violent. In these cases, the male was clearly the perpetrator of violent behavior. These forceful copulations MacKinnon has called "rape."* In Rijksen's study of Sumatran orangs, he identifies two basic patterns of sexual behavior: "rape" and cooperative mating. In Rijksen's own words:

> In its extreme form, the "rape" is a straight forward copulatory act, initiated by subadult males, in which females usually display distress and a lack of cooperation. In

* Even the newspapers have helped to tarnish the reputation of male orangs, as the following excerpt will demonstrate:

DEAR ANN LANDERS: I am twelve years old and very inquisitive about something. Do any male animals rape females of their species or is this something only humans do? I know this is a dumb question but I'd like an answer and can't find it in any of the encyclopedias. Thanks for your trouble. — Over-Active Mind

DEAR MIND: Male orangutans are rapists. And don't worry about your questions being dumb. The best way to learn is to ask.

its most extreme form, "cooperative mating" is an interaction, usually initiated by the female, in which both partners show considerable coordination of movements before, during and after the copulation, while copulation itself is performed cooperatively (p. 266).

Rijksen goes on to point out, however, that in many instances it is difficult to determine the proper category, since the degree of cooperation is often not at all clear. The female may avoid the advances of the male, but these efforts seem often to be half-hearted. We have observed the same behavior in many of our captive females. As an example of the difficulty of acquiring data on copulations in the wild, Rijksen observed 58 copulations in the Ketambe study, only five of which involved a pair of wild orangs. In 36 cases, the participants were a wild and rehabilitant orang, and in 17 interactions the pair was composed of 2 rehabilitants. It is clearly demonstrated here that only habituated or partially tame animals can be observed without environmental obstruction or reticence on the part of the subjects.

As Rijksen points out, all 5 observations of wild orang copulations were examples of "rape," as were 22 of the 36 wild-rehabilitant copulations. Interestingly, in all of the cases, the perpetrating male was a subadult. Because subadult males appear to be especially prone to

Table 4-1. Total number of observed combinations of sexually mature males and females and their sexual interactions.

Combination male	Combination female	Number of combinations	Number of consortship combinations	'Rapes'	Total number of copulations (incl. rapes)
M	Fi	23	–	–	–
M	F(j)	27	2	–	–
M	Af	23	2	–	–
S	Fi	25	1	1	1
S	F(j)	10	3	4	4
S	Af	41	–	–	–
M	RAf	(52)	1	–	–
S	RAf	(183)	2	6	17
RS	Af	(176)	1	11	19
RS	RAf	many	1	5	17

M: wild adult male; S: wild sub-adult male; RS: rehabilitant sub-adult male; Fi: wild female-infant unit; F(j): wild female-juvenile unit and lone potentially receptive female; Af: adolescent female; RAf: rehabilitant adolescent female. (After Rijksen, 1978)

forceful copulations in the wild, females apparently avoid subadult males when possible; especially receptive females (lone or with juvenile offspring). According to Rijksen, receptive females are the more likely targets of the subadult male, a strategy which makes perfect sense if these males are to effectively contribute to the gene pool. It also makes perfect sense for the receptive female to avoid these males, since the fully adult and successful adult male may be the more fit partner.* Contrary to this view, however, is the view of MacKinnon (1974) who believes that, at least in Bornean orangs, the younger males sire more offspring than the male "territory" holders.

According to MacKinnon, there are two phases of reproductive strategy employed by male orang-utans. The first phase begins at about age 10 and continues to about 15. In this first phase (*subadult sexuality*), the young male's strategy is to maintain consortships with relatively cooperative females, and forcefully copulate with uncooperative females. In the second phase, which MacKinnon termed the phase of *full adult sexuality,* the male is able to attract females, employ greater selectivity, and consort with a greater certainty of successful fertilization.

There are also two phases of agonistic activity according to MacKinnon's schema. The first phase (beginning at about 15 years of age) is that of the *full adult* in which the male begins to display and call. Presumably, this reduces the reproductive success of other adult males within the resident male's range. The second phase (*old adult*) commences in his early 20's and is characterized by a continued high degree of display and calling and reduced sexual activity. Presumably, this latter phase conveys an advantage for the male's offspring which benefit from the stablized home range. As MacKinnon concluded:

---

* This view was also suggested by Galdikas-Brindamour:

The evidence was piling up that competition between males for females was an important factor in orangutan adaptation. All the adolescent females we met preferred large mature males as sexual partners to the smaller subadult males who were their more frequent companions. (1975, p. 462)

The tendency of orang females to select adult males suggests a purpose to the distinct morphological features of adult males. Darwin's (1872) theory of sexual selection permits the argument that the secondary sexual characteristics of huge cheek pads, flabby throat pouches, large size, and hairy bodies all contribute to the visual complexity and presumed attractivity of these well endowed males. Beauty is clearly in the eye of the beholder, but female orangs may well be discriminate in their selection of the better developed males. The throat sac of the orang-utan (present in females which do *not* long call) has been presumed to act as a resonator for the territorial long call. However, we have noticed no marked reduction of long-calling intensity as a result of open fistulas on males with draining throat sac infections. It is interesting to note that the Yerkes suggested that the air sac might function as a swimming bladder. Perhaps the sexual selection argument best explains its size. Some further thinking about the matter is quite obviously required.

Figure 4-1. Resistance to copulation by female *Sungei* with male *Bukit* at the Atlanta Zoological Park (T. Maple photos).

In this way he may guarantee that his offspring have sufficient space, and therefore food, to grow up and breed after him. This tactic may be his best way of promoting his own lineage (p. 271).

The application of MacKinnon's developmental model accounts for the observations of all field workers, and adequately explains the different behaviors observed in the various age and sex classes. Further observations will be required to increase the sample size and ultimately validate the model.

A typical "rape" sequence between male and female orang-utan may be constructed as follows:

A feeding female might stop feeding, sit down and wait when she spotted an advancing sub-adult male. When such a male entered the fruit tree, she might make a squeak vocalization and remain immobile, or move in a direction away from the male, uttering squeaks. In the latter case her locomotion was usually characterized as hesitant avoid. Thus, the female Daisy, approached by the sub-adult male Doba, responded by moving away some meters, but then stopped and actually presented

Figure 4-2. Resistance to copulation by female *Sungei* with male *Bukit* at the Atlanta Zoological Park (T. Maple photos).

> . . . During positioning the females usually uttered squeak vocalizations, which might grade into whimper vocalizations and even screaming during the copulation . . . after the completion of a rape both partners behaved rather indifferently toward each other (Rijksen, 1978).

In the wild, as in captivity, newly arriving females and familiar females from which males have been separated for a time, are invariably subjected to forceful copulation. Of particular interest in this respect is the following observation by Rijksen:

> . . .when two females with consorting sub-adult males, met during a temporary association, the males of each consortship pair (i.e. Doba and Bor) raped the female (i.e. Bin and Daisy) of the other (p. 268).

Rijksen further asserts that this episode demonstrates that it is not simply a "built up sexual urge" which causes a subadult male to "rape." In my opinion, this remarkable observation demonstrates an effect which we have intended to examine in the laboratory: the so-called *Coolidge Effect*. First demonstrated with rodents, the phenom-

Figure 4-3. Resistance to copulation by female *Sungei* with male *Bukit* at the Atlanta Zoological Park (T. Maple photos).

enon accounts for the acceleration of male sexual performance when unfamiliar females are introduced to a seemingly exhausted male. While no one has yet experimentally demonstrated this with anthropoid apes (Rijksen's observations notwithstanding), its validity in human primates seems especially clear.

COOPERATIVE SEXUAL BEHAVIOR

Sumatran orang-utans which cooperatively mated in the Ketambe study area, also exhibited other cooperative social behavior. Traveling together, these animals sometimes stayed together for months (Rijksen, 1978), an observation supported by MacKinnon who found Bornean orangs to be less likely to enter into prolonged "consortships" than were members of the Sumatran subspecies. Whether this finding represents a true subspecies difference, regional idiosyncracies, or sampling error, is anybody's guess, but I favor the latter two possibilities.

From a sketchy data base, Rijksen nonetheless convincingly argued that consortships initiated by subadult males are directed to mature females, whereas adult females tend to select mature males as consorts. Again, in both instances an older animal may well be a more fit companion, thereby enhancing the genetic fitness of future offspring and the perpetuation of one's own genetic line (cf. Barash, 1977; Wilson 1975).

In the mating behavior sequence which Rijksen calls "cooperative," more neutral elements such as nest building, play, and grooming occur than in "rape" sequences. Clearly, cooperative mating is less stressful for both partners, and may well suggest, in many instances, the element of familiarity. Rijksen has stated that male inspection of female genitalia always occurs during cooperative mating, but is often absent during forceful copulations. However, as the reader will see, genital inspection is a common behavior in all captive sexual interactions, including those which can be classified as forceful.

SEXUAL BEHAVIOR IN LABORATORY AND ZOO

Only a few studies of captive orang-utan sexual behavior have been published and these have emerged primarily from recent work at the Yerkes Primate Center and the Atlanta Zoological Park (cf. Nadler, 1977; Maple, Zucker, and Dennon, 1979; Maple, Wilson, Zucker and Wilson, 1978; Maple, 1979).

In Nadler's experimental study, pairs of orangs (one Bornean, three Sumatran) were tested during the females' intermenstrual periods. In daily tests where animals were each time introduced for a limited time period, Nadler determined that the male orang-utan was the primary initiator of sexual interactions, copulating forcefully on a daily basis. Nadler further determined that orang-utans copulated for prolonged periods of time (range 60–2760 sec; X = 900 sec.), much longer than either gorillas or chimpanzees. Coffey (1975), Dennon (1977), MacKinnon (1974) and Galdikas (1979)* have made

* As compared to these captive data, Galdikas has written that Bornean orangs in Kalimantan generally copulate face-to-face with the male generally hanging below the female. From her observations, thrusting was determined to last 3 to 17 minutes. Galdikas (1979, p. 206) is in agreement with our findings in asserting that copulations were "almost invariably preceded by the male's oral contact with the female's genitalia."

similar observations which essentially confirm these reported durations. Moreover, when pelvic thrusts were counted, Nadler found that on this dimension too, orang-utans were more vigorous than their anthropoid cousins, the gorilla and the chimpanzee.

At the Regent's Park Zoo in London, MacKinnon made the following observation:

> Under captive conditions, with no natural outlets for travel and foraging, these animals undoubtedly diverted more time and energy into their sexual activities than their wild counterparts. I was, nevertheless, greatly impressed by the relative peacefulness of the proceedings and the females' willing participation in, and obvious enjoyment of, the sex-play and mating. This was very different from the rapes I had witnessed in Borneo. The relationships between *Gambar* and *Bunty* and *Boy* and *Twiggy* were so similar to those of the consorting pairs I had met in the wild that I was convinced that it was within such stable consortships, rather than from rape, that most wild babies were conceived. (1974, p. 179)

FEMALE SEXUAL BEHAVIOR

Some basic data regarding the female orang-utan can be considered here, although complete information has not yet been acquired. In 1938, Schultz suggested that there was evidence for cyclic genital swelling in orang-utan females. However, a study published by Graham in 1970 could not confirm this finding. If there is swelling, doubtful as it seems, it would be of little use to scientist, caretaker, or male orang-utan, since the extensive hair covering would render detection impossible. With greater certainty, we can trust that the female orang-utan's menstrual cycle is 29–32 days in duration, with menstruation lasting 3 to 4 days.

In a recent publication, Beach (1976) distinguished between the terms *sexual receptivity* and *sexual proceptivity*. Beach operationally defined the former as "behavior exhibited by females in response to stimuli normally provided by conspecific males" (p. 125). He went on to assert that the minimum of receptive responses in any species comprises those female reactions which are necessary and sufficient for fertile copulation. Proceptive behavior, according to Beach, consists of *appetitive* behavior exhibited by females in response to stimuli received from males. Both receptive and proceptive behavior are species-specific and represent two arbitrary ends of a continuum of behavior, which taken together comprise the female's sexual repertoire.

While it has been the task of animal behaviorists to completely describe the sexual repertoire of a given species, the female's role in copulation has frequently been ignored, distorted, or inadequately described. Moreover, it is often difficult to distinguish receptive from proceptive behaviors. For example, female gorillas and monkeys often play a more active role in copulation by lowering themselves onto a recumbent male's penis (gorilla) or reaching backwards to pull a mounted male closer to them (monkeys) (Beach, 1976). Both gorilla and chimpanzee females assume a posture which is known as a *present*. In this position, the female readies herself and presumably encourages the male to mount her. She accomplishes this by turning her body away from the male, pointing her protruding genitalia in the direction of his face. This posture is often accompanied by backward movement toward the partner, looking backward, backward hand extension, and several situation-specific expressions and vocalizations. It is not clear whether orang-utan females emit a present posture. Given the typical male's behavior, it may be an unnecessary and, therefore, vestigial behavior. In no sexual situation, including proceptive periods, (see also Chapter 4) have we observed presenting by any of our females. However, Rijksen reported that, during both cooperative and forceful copulations, females presented. In the latter instances, the females may have been engaging in appeasement signals, another context for present which occurs in other apes and in old world monkeys. To determine the incidence, context, and meaning of present for *Pongo* will require more research. For now we can conclude that present is at least a *rare* behavior for orangs, and in this sense the female sexual repertoire differs from that of *Pan* and *Gorilla*.

The orang-utan present described by Rijksen bears scant resemblance to that of chimps or gorillas, as the female is said to hang by two limbs with her legs spread apart and close to the male's face. This mode of present posture would of course be consistent with the arboreal propensities of orangs. Rijksen also noted that many females repeatedly scratched themselves when presenting. On one occasion, a female was seen in a posture resembling the "crouch present" of chimpanzees, but the typical orang-utan crouch resembles the "ventro-ventral present" of chimps according to Rijksen (cf. Van Lawick-Goodall, 1968).

In past descriptions of female sexual behavior, characteristics such as these were obscured by the general impression of female passivity in comparison with the conspicuous dominance of the males. In the great apes, the chimpanzee has been the subject of considerable attention, whereas the gorilla and orang-utan have been studied to a considerably lesser extent. Studies of the chimpanzee indicate that copulations, under natural conditions, occur only during midcycle when the female is presumably ovulating (McGinnis, 1973). At such times, chimpanzee females approach males, present their genitals to them, and vocalize to a greater degree than during any other stage of the cycle. In the gorilla, the female's initiative is even more conspicuous. Female gorillas are more assertive during midcycle, pursuing, pushing, and rubbing against the male's genitals (Hess, 1973; Nadler, 1976a).

In considering the active female role in chimps and gorillas, it is surprising that researchers have failed to conclusively demonstrate proceptivity in the orang-utan. As Nadler (1976b) pointed out, most accounts of the sexual behavior of captive orang-utans report frequent copulations independent of the female's cycle phase (cf. Fox, 1929; Asano, 1967; Heinrichs and Dillingham, 1970; Coffey, 1972). Similarly, field workers (cf. MacKinnon, 1974; Rijksen, 1975) have reported that males initiate copulation without regard to female cycle phase. In these encounters males may persist despite resistance by the females. One writer, however, reported some evidence of a cyclic female receptive phase wherein mating occurred for only 7 to 10 days in each twenty-five day period (Chaffee, 1967). Moreover, field workers have also suggested that the attraction of females to male long-calls may be related to her hormonal status (MacKinnon, 1968; Horr, 1972; Rodman, 1973; Rijksen, 1975; Galdikas, personal communication). In Nadler's (1976b) research, males copulated with females throughout all phases of the menstrual cycle during daily one-hour pairings. However, tests extended to five or six hours revealed that multiple copulations were more prevalent during the midcycle phase than at any other time. Thus, the evidence is equivocal and, while orang copulations may be subject to some degree of hormonal regulation, neither Chaffee nor Nadler distinguished between receptivity or proceptivity in females.* In our laboratory we have found such evidence, however, in an adult pair of orang-utans living in a fairly spacious habitat, with 24-hour access to each other.

The subjects of this study were two twenty-year old Sumatran orang-utans (*Pongo pygmaeus abelii*). The male, *Bukit*, and the female, *Sibu*, were feral born and captured as infants, arriving at the Yerkes Primate Center in 1962. During the course of the study, both animals resided in a 5.5 × 3.0 × 7.3 m. enclosure at the Grant Park Zoo in Atlanta, Georgia, on loan from the Yerkes Primate Research Center. The enclosure contained a series of intersecting welded metal pipes which allowed brachiation and climbing. Four elevated benches permitted arboreal sitting and sleeping, and the barred ceiling was utilized for suspension and locomotion.

Bukit and *Sibu* were introduced on June 25, 1976, and observed daily thereafter. Observations lasted for one to two hours, during which time the observer sat within two meters of the enclosure, recording by hand, the frequency, duration and sequence of behaviors using the ethogram described in chapter two.

From this study, the data concerned the number of times *Sibu* or *Bukit* initiated or attempted to initiate sexual behavior (mounts) and a descriptive account of the female's proceptive repertoire. Mounts were determined by the initiation of genital contact. In the male initiated behavior, this was obvious. A male "attempted" mount was defined as a pre-copulatory chase, hand-genital contact, or close visual inspection of the female's genitalia without copulation ensuing. Female-initiated mounts have been defined as sitting or thrusting upon the male genitalia. Unsuccessful ("attempted") mounts were those in which the female attempted to mount but the male moved away before the mount actually occurred.

MOUNTING

Figure 4-4 presents the number of mounts, either attempted or successful, for the first three months (92 days) of cohabitation. Female-initiated sexual behavior extended over a 4-to 6-day period, and ap-

* In his recent work Nadler has found some evidence of proceptive behavior and has designed some clever experiments—now in progress—to corraborate its presence.

peared 26–30 days apart. The last third of Figure 4-4 reveals a high point on the 80th day with erratic, but high scores for succeeding days. The male's overall sexual behavior occurred at a low, steady rate. For the first two months together, the male-initiated copulations peaked slightly after the female's proceptive period, while for the third month, increased male activity preceded the female's peak.

Another behavior indicating sexual receptivity was proximity to the male. Proximity was defined as the occupation of space within one body width of the other animal. Proximity, a confounded measure since copulation necessitates proximity, mirrored the general shape of Figure 4-4; peaking corresponding to the peaks in female mounting. Associated with frequent proximity was an increased amount of social grooming of the male by the female.

Other behaviors have been identified which are associated with the onset of female mounting (Maple *et al.*, 1979). Typically, a few days prior to and on the first day of mounting, the female's activity level began to increase, especially solitary play activity. She was observed to belly-slide on the floor, turn forward somersaults, swing vigorously by her feet while slapping the ground, and repeatedly stomp in the water trough with her feet and splash water with her hand. Other behaviors included increased brachiation and chain pulling. These behaviors were succeeded by following the male, slapping at the male while hanging from the bars and chain in a sloth-like posture,

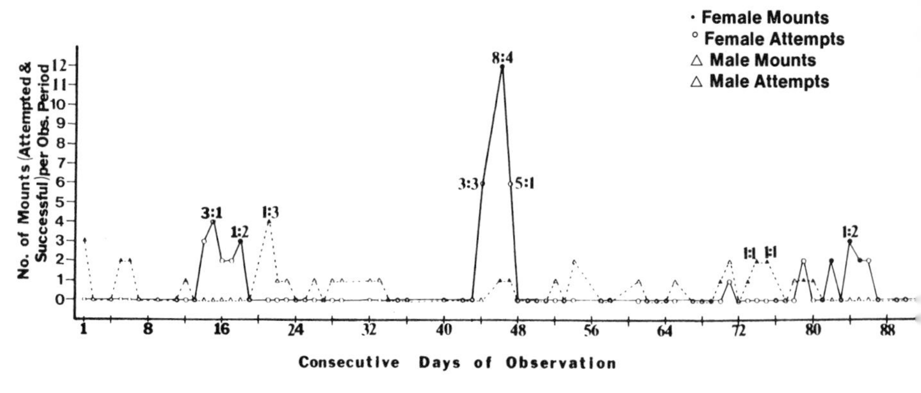

Figure 4-4. Number of mounts, attempted and successful, for three months of cohabitation of *Bukit* and *Sibu*. Ratios represent number of attempted mounts to number of successful mounts when both occurred on the same day.

and a vigorous rough-and-tumble play bout. These playful interactions occasionally led to male-initiated copulations.

Female-initiated mounts, unlike the play sequence, were comprised of attempts to induce copulations by the male. At these times, *Sibu* was observed to approach the male, stare at him, turn his shoulder, touch his genitals, roll him onto his back, part the hair around his genitals, position his penis, and then dorsally or ventrally sit and thrust upon it. Remarkably, *Bukit* rarely emitted reciprocal thrusts and usually withdrew from contact after several thrusts by *Sibu.* We never observed female mounting to result in male ejaculation. Galdikas (1979) has described an event similar to this as follows:

> This consortship differed radically from the ones involving adult pairs, in that it was clearly the adolescent female who was doing the following and often initiating copulation—sometimes only to be rebuffed. Indeed, at one point she even groomed him (something that rarely occurs among orangutans). (p. 217).

Galdikas later observed another female which examined a mature male's genitals, groomed him, and later sucked on his genitalia with no result. Four days later, after this female groomed his perineal region, he finally copulated with her.

On one notable occasion, we observed a variation in the sequence in which *Sibu* followed *Bukit* to a corner where he stood bipedally, grasping the cage bars with his hands. In this position, *Sibu* gained genital access by grasping his pelvis with her feet while she laid with shoulders, head, and arms on floor. In this position, she thrusted repeatedly against his penis.

In all female mounts, either attempted or successful, we observed persistence on the part of the female and reluctance on the part of the male. We have, for example, observed the female to fall onto her back and hold onto the male's arm, dragging behind as the male moves away. Interestingly, in only one of the eight sexual bouts observed by MacKinnon (1974) did the femal initiate contact with a male, and this advance was also ignored.

We have also identified a pursed-lip expression (cf. Hess, 1973, p. 561, for a discussion of the gorilla equivalent) which occured in the female only during female-mounting. The male occasionally emitted such an expression when he was engaged in copulation. The expression may be indicative of intense sexual stimulation or ecstasy (cf. Chevalier-Skolnikoff, 1974).

The finding of female-initiated sexual behavior was unexpected, as much of the literature had indicated a general pattern of male assertiveness as related to sexual activity. In fact, much has recently been written about the male orang's propensity for *rape* (cf. MacKinnon, 1974). Given this male propensity, cyclic changes in female behavior may be obscured in captivity and in the wild. In limited, daily access tests, as in Nadler's (1976b) study, male aggressiveness and limited space are likely to obfuscate the expression of any proceptive behavior by the female. Morever, hormonal fluctuations which may affect sexual behavior will be suppressed by pregnancy, (a likely and speedy result of male sexual vigor) creating a brief window during which time proceptivity may be recognized. With this in mind, it is perhaps *not* so surprising that no detailed descriptions of orang-utan proceptivity have previously appeared in the literature, and only few writers have even hinted at its likelihood (cf. MacKinnon, 1977).

From our research, we have presented data which indicate that female orang-utans may exhibit cyclic phases of sexual interest in males, culminating in attempts to copulate. We do not yet know that these bursts are coincident with the onset of ovulation. Menstruation is difficult to detect in the orang, and there are no visible physical signs of estrus in this species. Under more controlled conditions, physiological confirmation may be possible. However, in our study no proceptive behaviors were later observed after the three month period, and pregnancy was confirmed by veterinary examination in December. The estimated progress of prenatal development at that time (90 days) supported the contention that proceptivity correlated with the onset of ovulation.

Although we have observed proceptive behavior in only one female, we do not believe that it is idiosyncratic to her. Despite the differences in social organization of orang-utans, it would be surprising if female orang-utans did not exhibit some cyclicity in their sexual behavior, as has been reported for virtually all nonhuman primates. Our research of captive orangs suggests that wild females may seek out males during ovulation. Such a finding is suggested by differing degrees of receptivity as reported by Horr (1972), MacKinnon (1977), Rijksen (1975), and proceptivity as recently suggested by Galdikas (1979). One technique which might be successfully used in future studies would be radio-telemetry tracking of selected animals.

By this method cyclic patterns of travel towards a male's territory by a female may become discernable.* In captivity, the male's sexual behavior pattern corresponds to that of the free-ranging state: males are relatively free of hormonal regulation. Further research in natural or semi-natural habitats is needed to corroborate the findings that female-initiated proceptive behavior in captivity has a counterpart in the wild state, (Nadler, 1974).

PLAY AND ADULT SEXUAL BEHAVIOR

Play behavior has been observed between members of a consort pair, occasionally preceding copulation (see also Chapter 2). In the wild, one male was observed by MacKinnon (1974) on three occasions to initiate sex-play which led to copulation. It is possible that social play, as well as strengthening social bonds between young animals, may also be an initial behavior in the copulation sequence and, as I have suggested, a mode of female proceptive behavior.

In captive settings, copulation sequences generally begin with a precopulatory chase of the female by the male, with the female resisting the male's advances. Following this, the female submits or is wrestled down by the male and generally lies on her back on the floor. The male then approaches her, frequently spreads her legs, may engage in close visual inspection of her genitals and a remaneuvering of her position, and then mates with her often in a crouching or sitting posture (Coffey, 1975; Fox, 1929; MacKinnon, 1974a, Nadler, 1976; Zucker, Brogdon and Maple, 1976). Although the predominant position during copulation has been reported as ventroventral, many varied positions and postures have been observed, including dorso-ventral, with the female either lying on her stomach or standing quadrupedally (Coffey, 1975; Zucker, Brogdon and Maple, 1976). The male often supports himself by holding onto bars. In fact, wherever there are structures to grasp, both partners will do so during copulation bouts. Copulations may occur in suspended positions as well as on the floor, though some authors cite the latter as the most common in captivity (Coffey, 1975; MacKinnon, 1974a). Copulation duration may also vary, but usually lasts approximately 15

* Rain-forest conditions, however, may render this techique very difficult.

minutes (Coffey, 1975; Nadler, 1976). As previously discussed, Nadler (1976) has found that the number of pelvic thrusts by the male is greatest during the midcycle period of the female. In the field, MacKinnon (1974a) has reported high pitched rhythmical cries emitted by the female during copulation. He also described a male display of posturing in which the male stood bipedally and extended his arms to branch supports. This posturing, combined with swaying, was seen on one occasion when a male was initiating consortship with a female.

In reviewing the literature on copulation, an interesting hypothesis becomes apparent. It is possible that the male-initiated precopulatory chase serves as an arousal function for the male and may be a necessary element for successful copulation (cf. Zucker, Brogdon and Maple, 1976). Although Beach (1976) proposed that females which actively "solicit" copulation are more attractive to males and have a higher stimulus value than those which do not engage in these proceptive behaviors, observations of orang-utans seem to conflict with this. At the Atlanta Zoo, as we have seen, one captive male's response to the proceptive behaviors of the female was essentially one of indifference, although successful copulation did occur during this period. Two captive groups observed by MacKinnon (1974b), in which a male and two females were caged together, showed similar patterns. In both groups, one female actively approached the male and sought his attention. However, both males preferred to solicit the attention of the uninterested females which avoided them and even aggressively fought them off. At the Atlanta Zoo, on two occasions when a captive female was observed to submit to a male without a chase, he walked away (Zucker, Brogdon and Maple, 1976). The chase itself may not be aggressive or fear evoking in the females, for observers have noted that after the initial chase the female passively submits and assumes a position on her back which would seem to invite copulation (Fox, 1929; MacKinnon, 1974a,b; Nadler, 1976; Zucker, Brogdon and Maple, 1976). As MacKinnon has interpreted it, a male in the wild was observed to tickle and bite his consort female in an attempt to encourage her to participate in more bouts of struggling and chasing (MacKinnon, 1974b).

Returning to Beach (1976), proximity initation and maintenance by the female are examples of affiliative behavior and may be used to indicate when the female is in estrus. A strong cyclic pattern in proximity initiation by the female was also recorded for the pair *Bukit-Sibu*. The female exhibited an increased tendency to initiate proximity with the male on days 17–20, 47–51, and 83–90, becoming pregnant between days 73 and 103. After this time, cycling in proximity initiation ceased and the frequency of female-initiated proximity decreased. These cycling trends can also be seen in the duration of female-initiated proximities, however the trend was not as dramatic as it was for frequency scores. The higher peaks for frequency may be understood in terms of the alternating approach and withdrawal behavior of the female described by Beach (1976) as characteristic of proceptivity. These frequent approaches would not necessarily be reflected in a large increase in duration.

The female of a second pair, observed at Yerkes, *Durian-Paddi*, exhibited a similar trend in proximity initiation. This female increased proximity initiation and duration on days 10 and days 42-45. The second peak is not as large as the first with respect to frequency. However, this may reflect the fact that during the period of the second peak the male showed high proximity initiation. There may have been no need for the female to initiate proximity frequently.

The second pair at the zoo, *Lipis-Sungei*, did not show any cyclic peaks in female-initated proximity, due to the female's pregnancy. The female did exhibit higher frequency of proximity initiation on several days however. This may be due to the length and amicable nature of this pair's relationship.

At Yerkes, although the female *Ini* did not show cyclic proximity initiation, the frequency of this behavior was relatively high over the period from day 1 to 29. After this time, proximity initiation was depressed. The observer was informed of this animal's pregnancy on day 29 of the observation period. This pregnancy may explain the depression in proximity initiation.

The female *Lada*, when paired with *Bagan*, did not exhibit *cyclic* peaks in proximity initiation. Only one peak was observed on days 35–36. However, in the latter half of the study, proximity initiation by this female remained at a high frequency, with only slight fluctuations. This reflects high incidence of social play.

Similarly, *Paddi* did not exhibit cyclic peaks when paired with the male *Padang*, although she did initiate proximity to him on occasion.

When paired with *Sampit*, *Lada* initiated no proximity. Males did not exhibit *any* cyclic trends in proximity initiation in the Dennon and Maple study.

Since play occasionally serves as an initial behavior in a copulatory sequence, it may also be an indicator of female proceptivity (Maple, Zucker and Dennon, 1979). In the Dennon/Maple study, solitary play was observed to directly precede copulation on only one occasion. Therefore no strong support was obtained for MacKinnon's hypothesis. Play does appear to be cyclic for two females however. One of these was *Sibu* (the animal studied in Maple, *et al.*, 1979). This cyclic play behavior consisted of gradual increases and decreases in its frequency of occurrence rather than sharp, well defined peaks. These periods of increased play overlapped the periods of increased activity for the other postulated estrus-indicating behaviors.

Sibu increased solitary play on days 12–32, 46–65 and 73–94. Solitary play was not observed after this time. This absence of solitary play coincided with the period of the female's pregnancy. The male, *Bukit*, showed interest in her activity periods and frequently approached while she was playing. Social play occurred following male approach on three occasions. Moreover, the male was observed to initiate a chase sequence after one solitary play bout by the female, and to subsequently engage in ventro-ventral copulation.

Lada, when paired with *Bagan*, also evinced a cyclic trend in the occurrence of solitary play. This behavior occurred from days 21–44, 57–70 and 79–94. Again, these increases in solitary play behavior coincided with the period of increased social grooming exhibited by this female. The male in this pair did not appear to take interest in these female activities.

The final two females to exhibit solitary play did so infrequently. It was not possible, therefore, to determine if these females were cyclic in this behavior.

Female-initiated social play was cyclic for *Sibu* and social play was observed on day 16, 46, 51 and 61, and days 73–93, with a concentration of social play on days 79-82. These peaks of play activity correspond to the periods of proceptivity inferred from other behavioral measures for this female. Social play was not observed after this female became pregnant. It is possible that social play may be used as an additional indicator of receptive periods in certain females.

The second female to show frequent social play behavior was *Lada*, when paired with *Bagan*. This female did not show cyclic trends in the occurrence of this behavior, however. Instead, she played socially on day 42, reaching a high frequency of play from day 56 to the end of the study. It may be possible to explain this lack of cyclicity in social play by reference to the environment in which this pair was studied. This pair was housed at the Yerkes Primate Center and observed in either the inside or outside cages. The inside cage was smaller, provided less room for locomotion, and did not permit visual access to other animals. Hence, it may have contributed to the result.

Male Sexual Behavior in Captivity

Forceful "raping" of the female by the male, and peaceful consortships between males and females were both seen in the Maple and Dennon study. These types were not exclusive of each other within a given pair. Rather, all pairs revealed a mixture of both types. While MacKinnon (1974b) suggested that a difference between Bornean and Sumatran orang-utans is the propensity to "rape," both types of copulation were recorded for the pairs of each subspecies in our study.

All but five of the male-initiated copulations involved a chase component prior to copulation. One male (*Durian*) was observed to withdraw from copulation on three occasions, only to chase the female again as soon as she withdrew, and then resume copulating with her. On one of these occasions, he was observed to push the female away before resuming the chase. Again, this is similar to an observation in the wild in which a male was seen to encourage a female to participate in additional bouts of chase (MacKinnon, 1974b). Female-initiated copulation attempts, which lacked a chase component, generally met with little male response. These findings provide support for the idea that a chase element is arousing to the male. A comparison of the length of copulation bouts with respect to the presence or absence of a chase component revealed that copulations preceded by a chase were significantly *longer* than those which were not initiated by a chase—11 minutes, 34 seconds compared to 2 minutes, 23 seconds ($t = 4.348$, $df = 33$, $p < .001$). This difference in copulation length may also reflect a difference in the success of copulation due to the presence or absence of a chase component.

This chase element is generally characterized as being aerial,* involving several circuits of the animals' cage. Considering this aerial component, it is interesting to note the difference in chase length for the Atlanta zoo animals as compared to those housed on the Great Ape Wing of the Yerkes Primate Center. The Yerkes animals were in much more restricted living quarters with fewer aerial pathways than the animals housed at the zoo. Chase length at the zoo averaged 2 minutes, 14 seconds whereas at the Yerkes Center it averaged only 37 seconds. This difference was significant ($t = 5.67$, $df = 23$, $p < .001$). Also worth noting is the fact that the five male-initiated copulations which occurred without the chase component took place at Yerkes.

This chase component is not necessarily aggressive nor does it seem to induce fear in the female. Rather, several chases ended by a female assuming a copulatory position on the floor. Once copulation was initiated, the two primary postures were observed as mentioned earlier. These appeared to be generated as a function of the female's behavior and the male's response to it. Aggression was recorded for those copulations in which the female was uncooperative only. This consisted of the female squirming, trying to escape, or otherwise avoid the male. Males generally responded to this behavior by biting the female on the hand or foot, or threatening her. Sixteen cases of uncooperative copulation were observed in the Dennon and Maple study and only one of these did not involve an aggressive response by a male. Aggressive copulations did not have any clear relationship to inferred female receptivity. Rather these were intermixed with passive bouts throughout high frequency copulation periods. Further research in this area however, might reveal a relationship between these two elements, e.g. in the duration of female struggling and her hormonal status.

---

* The aerial component of the copulation behavior described in the present study corresponds to the arboreal chases preceding this behavior in the wild. This behavior has been referred to as an *arboreal* chase by other authors. The term aerial is used here as a more accurate description considering the treeless nature of these animals' habitat in captivity. The same behavioral element is inferred.

On three occasions females responded to a male's aggression by becoming passive for the remainder of the copulation. A male was observed to aggress against a passive female on only one occasion. Female passivity consisted of a female assuming a position on her back, often prior to being reached after the chase, and remaining passive throughout copulation. Observations of this behavior support findings by several authors (Fox, 1929; MacKinnon, 1974a, b; Nadler, 1976; Zucker, Brogdon and Maple, 1976).

The rough turning and shoving of the female, and biting of her hands and feet is a complex of behaviors which I have called *enforcement**. While turning** and manipulating the female may be characteristic of orangs, I have chosen this term to signify hitting and biting in a sexual context. Activity such as this may be related to the males apparent difficulty in maintaining intromission. Orangutans are extremely hairy creatures, especially the males, which also have rather small instruments for penetration. Consequently, males are continually adjusting themselves and their partner's posture. They also engage in considerable *furplay*, manipulating the females vaginal region with their hands and mouth. Pelvic thrusting is facilitated by the Orang's prehensile feet which are often employed in grasping the female, particularly during aerial copulations. Enforcement appears to be related to the frustration of *coitus interruptus*, where the female moves incorrectly, or in some other fashion interferes with sustained intromission. Aggression is particularly prominent where orangs are introduced for the first time, after varying periods of separation, and/or in a small test cage. Aggression occurs in our pair-living zoo males, but it is less intense.

From seven recorded bouts of adult sexual behavior in the wild, MacKinnon (1979) observed that males typically clasped females at the waist or thighs with their feet, thrusting ventrally or from the rear. The copulations observed by MacKinnon lasted "about ten minutes." In four of these instances, males struck and bit the

---

* The hands and feet are often bitten in the play of juveniles (Clifton, 1976) and adults (Maple and Zucker, 1978).

** The male orang *Lipis* engaged in much turning of the female *Sungei*, particularly after the birth of her infant. I believe that *Lipis* prefers the d-v position, since he does not turn her once she has assumed this posture. She persists in initially assuming the v-v position, however, which is a particular likelihood when ventrally carrying an infant. The d-v position may be a more efficient position for this male.

struggling females (p. 261). During many copulations, females were observed to emit a rhythmic squeal vocalization.

As we have seen, male-initiated copulations are characteristically long, approaching an hour in some instances. A male's ejaculation is not always easy to detect, but it is characterized by a rigidity of the body, a lengthy pause of 10–15 seconds (but we also observed frequent lengthy pauses throughout copulation), and a relaxation of the body upon the quick withdrawal of the female. In fact, the female's behavior, we believe, is a potentially useful indication of male ejaculation. After ejaculation, the males examine their genitals, eat excess ejaculate, and typically move away from the female, often to drink water. We have observed no vocalizations during copulations of orangs that are living continuously together, although other investigators have reported both male and female vocalizations.

MALE COMPETITION IN CAPTIVITY

One example of male competition has been recorded in several American zoos, but has remained a part of the unwritten zoo lore. In these instances, a fully adult male caged with a female and their male offspring coexists with no signs of aggression until the young male reaches the approximate age of four. In one such case at the Atlanta Zoo, the male *Lipis* severly injured the female *Sibu* while fighting with her and the four-year-old male *Lunak*. We have hypothesized that the male of four years or older is strong enough to interfere with the older male's sexual advances toward the female.* Because captive four-year-olds are still dependent, their interference can trigger aggression. As we will see in Chapter 5, adult males are extremely serious about sex, and are known to bite and hit their consorts if they fail to cooperate. Similar altercations between adults and four-year-old males have led me to suggest that this is the age at which young males should be separated. The time is consistent with the increasing peripheralization of young males in the wild as they move into favorable territories. Nadler and Braggio (1974) have provided some evi-

* This view is in line with an observation of Galdikas (1979) who reported that during a forceful copulation, the large infant of the adult female *Beth* participated in the struggle against the subadult male *Mute* (p. 208). MacKinnon also observed infant participation in resisting the males attempts to forcefully copulate with their mothers (1979, p. 261).

dence to support this in their discovery that even younger males exhibit signs of territoriality and peripheralization in their play patterns. Rijksen (1978) has also discussed the developing adult male in Sumatra.

> . . . a dependent orang-utan youngster becomes thoroughly and exclusively familiar with his mother's core area for the first four years of his life. As a result of his acquaintanceship with peers from adjacent core areas and his participation in peer groups, the growing adolescent extends his range beyond that of his mother's core area. This process probably continues well into sub-adulthood until he eventually has become familiar with quite a number of residents. Within this area the late sub-adult male may attempt to settle into a sector where the social and other conditions are favourable. It may be that there is considerable competition for such areas. (p. 165)

An even more remarkable example of zoo lore is that where young males are not removed from an enclosure inhabited by a fully adult male, the complete development of secondary sexual characteristics seems to be suppressed. I have observed this situation in several American zoos and at the Yerkes Primate Center. In the latter case, the animal developed full cheek pads within a few months after his separation from the dominant cage mate. An interesting example of this phenomenon took place in a European zoo where an adult male orang was subordinate to the primate keeper. The orang only developed his cheek pads after successfully challenging his human caretaker (Schmidt, personal communication). This phenomenon may be a result of low testosterone output which accompanies subordination. The function of the suppressed structural development may permit male orangs to appease older males until such time that they can acquire a territory of their own. I should stress here that this observation is a part of the lore, and hard data have not been acquired either to substantiate its occurrence or support my hypothesis about its function. Further research on this fascinating phenomenon is obviously necessary.

UNUSUAL MALE SEXUAL BEHAVIOR

In captivity, adult male orang-utans which live alone are prone to frequent masturbation. In fact, where masturbation is absent, the ability of the animal to breed may be suspect. Male orangs at the Yerkes Primate Center have been known to masturbate by thrusting into the holes of chain-link fencing for stimulation. This behavior is especially prominent if females are housed next door. At the Sacramento (California) Zoological Park, an adult male was observed as he manually induced his female cage-mate to stroke his penis with her hand (cf. Maple, 1977).

Homosexual behavior among adult males has also been observed in captivity. At Yerkes, two males (*Dinding* and *Durian*) regularly mouthed the penis of the other on a reciprocal basis. This behavior, however, may be nutritively rather than sexually motivated. We have also observed it in the juvenile male *Lok-lok*.

Finally, orang-utan males have been known to become aroused at the birth of an infant (cf. Mitchell, 1979). As Harrisson described it for an event occurring at the Philadelphia Zoo:

> This is a danger common at zoos: conditions of overcrowding. When Guarina gave birth to her seventh baby, Pinky, her mate Guas was allowed to remain with her during the entire pregnancy and birth. The birth excited Guas sexually and he attempted to copulate with Guarina, who tried to avoid him. In the ensuing scuffle the baby was treated very roughtly and the father had to be removed. However, the baby survived the ordeal. (1962, p. 145)

This propensity does not seem to occur very often, but it has become prominently mentioned in the zoo lore. As a result, zoo staff have been reluctant to permit males access to females which are about to give birth. In fact, the potential sexual aggressiveness of males often leads to separation as early as the fifth month of pregnancy. Contrary to these fears, we have observed three different adult males present at the birth of a total of eight infants. On occasion, these males have exhibited interest in the placenta, but have generally remained disinterested in the offspring. Moreover, while we have observed occasional copulations in the latter stages of pregnancy, where space is adequate, females generally choose to evade the advances of males, and usually do so successfully. Adult males, present during birth,* *should* be carefully watched. After all, orang males are probably rarely, if ever, present at births in the wild. However, we believe that the danger of their presence has been overrated.

---

* See also chapter 5 for a further discussion of the problem.

In fact, the absence of a long-time cage mate may well adversely affect the birth due to the stress of separation. Moreover, when the male is later united with the mother-infant dyad, after prolonged social deprivation, he will most certainly be sexually (and aggressively) responsive.

FEMALE DOMINANCE IN CAPTIVITY

Dominance among captive females has been examined by Nadler and Tilford (1977). In this study four adult females (with infants) at the Yerkes Primate Center were pair-tested both for social dominance, and in competition for food. In the first test, social dominance was readily established whereby one animal repeatedly displaced another. In addition the socially dominant animal was the one that generally obtained the food incentives.

Dominant animals were characterized by the two primary behaviors, each of which exceeded in frequency those of the subordinate: hair pulling and examination of the other's infant. Subordinate animals defecated and urinated more frequently than did dominant animals, and remained suspended from the cage bars for a much longer period than did their dominant cagemates. An especially interesting result was that orang females were dominant regardless of which home cage was used for testing. Thus, in these particular females, there was no observable advantage due to "territory." In only one pair did social dominance fail to correlate with priority of access to food. This finding, however, suggests the potential complexity in the measurement of dominance relations. Finally, dominance was *not* related to size, since the *most* dominant animal was the *smallest* in weight. The most aggressive behavior, however, was emitted by the two heaviest females.

Although Nadler and Tilford suggested that female dominance interactions as observed in their captive study are probably rare in nature, it is the opinion of this writer that they may only be obscured in the wild due to the likely proximity of related older offspring and mothers (cf. Horr, 1973). Thus, the false impression of tolerance may be an artifact of species-typical social organization. Anyone who has observed orang groups in zoos is aware of the expressions of dominance among nonrelated females, including furious fighting over favored areas and favored foods. We have also observed instances of dominance reversals which were apparently the result of either the presence of a newborn infant (the animal *with* infant is dominant over one *without*), the presence of a control role consort bond (a familiar male provides a safe base from which to attack a lone female), and/or renewed competition for a male which was formerly available on an exclusive basis (the newly available, and formerly dominant, female is attacked while being sexually pursued by the male). Here again, dominance relations among captive orangs are complex affairs which cannot be easily predicted.

COPULATION POSTURES

The preferred copulation position in our Yerkes and Zoo studies was ventro-ventral, occurring in 67 of 118 cases or 57% of the time. This position most frequently involved the female lying on her back with the male sitting or leaning slightly forward over her. The female's assumption of a position on her back when she demonstrated receptive behaviors, may reflect her recognition of the ventro-ventral position as the one preferred by the male. Dorso-ventral copulation was the second most frequent position observed occurring 28% of the time or on 33 occasions. Latero-ventral copulation occurred on 18 occasions or 15% of the time. The males were observed to manuever the females on occasion, flipping them over into the desired copulation position. Frequently however, the dorso-ventral position was a result of the female's attempts to escape the male. Copulation occurred most frequently on the ground, 42 cases as compared to 10 cases of aerial or suspended copulation. In the wild, most copulations occur while in trees. An especially interesting early description was provided by Fox (1929) as follows:

> The copulatory act of our orangs is worthy of description because of its dissimilarity from that described for the chimpanzee. When the desire animates the male and is reciprocated by the female, he pushes and mauls her a little, whereupon she lies upon her back on the floor. The male then approaches her and separates her legs. During the act he remains in a sitting or crouching attitude and though they are face to face, he does not lie upon the abdomen of the female. The male will sometimes grasp a leg of the female and hold it up and to the side during conjugation. During the act, there is no fondling, nor do they mouth each other either before or after the

act. The female lies passive and often has a hand over her face. The act is practiced daily, without relation to the sexual cycle. The period and interval of the sexual cycle has not been definitely established here. It is thought that the very slight bloody flow lasts 3 or 4 days. The frequency of copulation in the chimpanzee is greater. The act in the latter is similar to that described as having occurred in Havana, although there is no knocking on the floor with the backs of the hands. At no time has the keeper observed in either animal that copulation has occurred *more canum*.

Genital inspection immediately preceded copulation on 19 occasions and occurred during the copulation bout on 17 occasions. Limb manipulation of the female by the male occurred in 16 copulation events. This consisted of the male moving the female's arms or legs while copulating with her. The legs of the female were frequently held up near her head. Often the female would be observed to hold her own legs in this position. It is believed that this positioning aided the male in his insertion. Another behavior which was interpreted in a similar fashion was frequently observed in some pairs during copulation. This behavior consisted of the male stroking the fur away from the female's genital region as well as his own prior to reinserting during copulation. The male was also observed to insert his thumb in the female's vagina and pull upward. These behaviors were also noted in two males at Yerkes.

Females ended more copulation bouts than did males. On 7 occasions the male withdrew, whereas the female withdrew on 25 occasions. Nine of those occasions on which the female withdrew however, were preceded by the male pulling out and ending the actual copulation immediately prior to her withdrawal. Copulation frequency and duration have been suggested by several authors to be indicators of the female receptivity (Chaffee, 1967; Goy and Resko, 1972; Nadler, 1976). In the present study only three males copulated often enough to examine the record for evidence of cyclicity. *Bukit* (one of the males at the Atlanta zoo) exhibited a cyclic pattern in both frequency and duration of copulation. Copulation was observed between days 8 and 25, 50, 74–84, 108–110, 154 and 163. The copulation occurring on day 50 was female-initiated. The first three peaks in copulation duration were associated in time with sexual proceptivity by the female. These periods were observed on days 17–25, 45–51, and 75–92. The latter days of male copulation occurred after the postulated female impregnation period.

The other two males exhibiting a high incidence of copulation were both members of pairs established during the course of the Dennon/Maple study. As a result, copulation was very frequent during the initial observation days. After this initial period however, *Durian* exhibited a high frequency of copulation on day 16 and again on day 42. Similar trends were seen in copulation duration scores. The female in this pair, *Paddi*, socially groomed on days 4 and 7, and again on day 38. Her proximity initiation with the male peaked on days 10 and 42. Therefore, male interest in copulation correlates to some extent with the female's periods of increased social behavior.

The third male, *Sampit*, also exhibited a high degree of copulation activity during the initial days of introduction to *Lada*. After this time, copulation duration peaked on days 9, 35 and 37. The female of this pair however, did not emit any social behavior toward the male during the course of the study. Although his behavior appeared to exhibit a cyclic trend, it cannot be conclusively associated with the receptive period of the female as data were not collected for this pair from days 21 to 35. The cyclic nature of this behavior may therefore be an artifact. Based on data from all three males, their interest in copulation may be cyclic to some extent and can therefore be postulated as a possible additional means of determining female sexual status.

MacKinnon (1974b) reported peaceful interactions between males and females in which the male inspected the female's genitals without following this inspection with copulation. Similar events were observed in the Dennon and Maple study. On eight occasions a male was observed to inspect a female's genitals without copulation, but only two males, *Bukit* and *Durian*, were observed to engage in such activity.

Also in the Dennon/Maple project, copulation occurred on 31 of 312 observation days, with a probability of .10 that copulation would be observed on any given day. The range of frequencies of observed copulations per pair was from 0 for *Dyak-Ini*, to 15 for *Bukit-Sibu*. One copulation bout was recorded for *Lipis-Sungei* and *Padang-Paddi*. Two copulations were observed for *Bagan-Lada*. Five and seven copulations were observed for *Sampit-Lada* and *Durian-Paddi*, respectively. The greatest probability that copulation would be observed on any given day was for the two newest established pairs—.44 for *Durian-Paddi* and .33 for *Sampit-Lada.* The probability that copulation would be observed in the other pairs ranged from .3 for *Padang-Paddi,* and .4 for both *Lipis-Sungei* and *Bagan-Lada.*

Copulations frequently occurred near feeding time, and reports from animal keepers indicated that copulation was frequently observed in the early morning prior to the first daily feeding. One instance of copulation in the Dennon/Maple study occurred between two attempts by the male to take the female's food, and copulation was once observed to follow food sharing in another pair. The chase sequence frequently seen in food sharing resembled the initial chase element of copulation. It is quite possible that there is an overlap between sexual arousal and hunger arousal in captivity.

Copulation was observed in the first pair, *Bukit-Sibu,* on 15 of 149 observation days. Seven of these copulation bouts contained aggressive elements. The lengths of these copulations were compared to those bouts in which the female was passive or cooperative. The results for this pair revealed a significant difference in the length of copulations with respect to the cooperative behavior of the female. Aggressive copulations averaged 17 minutes, 19 seconds, whereas copulations with passive females averaged only 9 minutes, 22 seconds ($t = 3.478$, $df = 13$, $p < .01$). These results support our proposed frustration-aggression interpretation of aggressive copulations.

Passive or cooperative behavior was recorded for the female of this pair on eight occasions. Two of these events were in response to overt aggression by the male during copulation. After being bitten or threatened by him, the female assumed a position on her back and remained there without struggling. Another form of passive behavior by the female occurred on three days. In this mode, the female ended a chase by assuming a position on her back. This behavior led to ventro-ventral copulations initiated by the male. On another occasion, the female was observed to stop fleeing from the male during a chase. This was followed by genital inspection by the male after which the female again avoided the male. In one instance, after the male caught the female, she was described as remaining passive throughout the bout while the male repositioned her and proceeded with the copulation. A final mode of female cooperation is copulation when initiated by the female. This behavior has been discussed in a previous section.

As mentioned earlier, ventro-ventral copulation was the preferred position of copulation for most pairs. In *Bukit-Sibu* this form was seen in 53% of the positions observed, with dorso-ventral copulation occurring 47% of the time. 49 positions were recorded for this pair. These scores do not accurately reflect the male's preference however, as many of the original dorso-ventral positions were the result of the female's attempts to withdraw from the male. Taking these into consideration, the male engaged in true dorso-ventral copulation only six times or 19% of the time. Therefore, the male preferred ventro-ventral copulation 81% of the time. With *Bukit* and *Sibu*, five cases of genital inspections occurred in which no copulation bout followed. One of these involved the male approaching the female, holding her legs and grooming her. The other four involved the male touching the female's genital region and then sniffing his finger and withdrawing. It is interesting to note that three of these events occurred after the female was impregnated. Copulation was not observed after day 163, although the male was observed to chase the female and engage in this type of genital inspection after this date.

Another pair worthy of detailed discussion is *Durian-Paddi.* Copulation was observed in this pair on 7 of 16 days. Five cases of aggressive copulation were observed. The length of copulation was compared with respect to this aggressive element, excluding those copulations in which no chase was recorded. The results revealed a trend in the expected direction; 5 minutes, 7 seconds for non-aggressive copulation as compared to 7 minutes, 16 seconds for aggressive copulations. This difference was not significant.

The female *Paddi* exhibited passive behavior on 7 days (16 separate occasions). Fifteen of these passive behaviors occurred when she ended a chase sequence by assuming a position on her back before the male reached her. On one occasion she assumed this position after the male pulled her to the floor.

This pair exhibited ventro-ventral copulation on 24 occasions, latero-ventral copulation on 10, and dorso-ventral on 4. However, all of the dorso-ventral positions were the result of female escape attempts, and one latero-ventral position was female-initiated. Therefore, considering this information, the male preferred ventro-ventral copulation 73% of the time and latero-ventral copulation 27% of the time.

In this pair, the female withdrew from copulation on seven occasions, only one of which was preceded by the male ending the actual copulation. The male withdrew from the female on six occasions. During several copulations, the male exhibited an unusual facial expression. His mouth was open, with the lower jaw loose. The lower lip was frequently relaxed and extended downward. The tongue was occasionally protruded. The average length of aggressive copulations was longer than that for nonaggressive copulation, but this difference was not significant—14 minutes, 38 seconds as compared to 11 minutes, 48 seconds. Female passivity was observed on two days. Each case involved the female assuming a position on her back to end the chase by the male. This was followed by ventro-ventral copulation by the male.

For *Sampit-Lada*, ventro-ventral copulation was observed 59% of the time. This compared with latero-ventral copulation 24% of the time and dorso-ventral copulation 18% of the time. However, two of the three instances of dorso-ventral copulation were the result of the female's escape attempts and one latero-ventral copulation was due to the female. Therefore, the male showed a preference for ventro-ventral copulation 71% of the time, dorso-ventral 7% of the time and latero-ventral 21% of the time.

The female withdrew from copulation on four occasions. Three of these were preceded by the male ending the copulation itself. The male was not observed to withdraw from any of the copulations.

Two copulations were observed with the pair *Bagan-Lada*. Although the female appeared to be restless in the first of these, the male did not show any aggressive behavior in response to this. This copulation lasted 23 minutes, 15 seconds, however. Of 12 positions observed, 42% were ventro-ventral, 25% were dorso-ventral and 33% were latero-ventral. However, the female initiated all cases of latero-ventral copulation, two of the three cases of dorso-ventral copulation, and one case of ventro-ventral copulation. Therefore, the revised scores for male preference are ventro-ventral copulation 80%, and dorso-ventral only 20%. Genital inspection without copulation was not observed in this pair.

Only one copulation event was recorded for the pair *Padang-Paddi*. This copulation contained aggressive elements in response to escape attempts by the female. In this case, the male was observed to punch the female in the face when she struggled. This V–V copulation lasted 6 minutes, 30 seconds, and *Paddi* withdrew after the male had ended the copulation itself. This copulation lasted 6 minutes, 30 seconds. The final pair, *Lipis-Sungei*, copulated only once. This bout occurred in between two attempts by the male to take the female's food. It occurred in the ventro-ventral position. Although no genital inspection was observed in this pair, they were observed to sit together with their arms around each other on two occasions. These results are similar to findings described by MacKinnon (1974a) for consort pairs which remained together for several months. The present pair displayed many social behaviors which reflected their long-term, well-established relationship.

It has been reported that compatibility between the members of a pair and sexual initiative by the female are important for breeding success in chimpanzees and gorillas (Nadler, 1976). This does not appear to be true for the orang-utan. The male may initiate copulation regardless of the behavior exhibited by the female. An excellent example of this may be seen in the behavior of *Sampit* when paired with *Lada*. Copulations occurred frequently even though this female exhibited no overt receptive or proceptive behaviors. Copulations were observed to occur in several pairs whether the female was passive or struggled to escape. In the present study, however, aggression by the male was only observed in those instances when the female did not cooperate. It is important to note however, that although matings may occur regardless of the compatibility of the pair, successful breedings may not be completely independent of this factor. Rearing the infant may be impaired and wounding of the female or the infant may occur as a result of incompatibility. An example of the harmful effects which may occur was seen during the time *Sibu* was housed with *Lipis* at the zoo. After being attacked by *Lipis*, *Sibu* aborted her pregnancy and showed overt fear reactions to the male after this time. These reactions were so intense that *Sibu* had to be removed from this social group.

The present studies of adults were carried out to extend the existing knowledge of the sexual and social behavior of the orang-utan. However, further work is needed in this area. Several studies should be carried out to fill in the gaps still present in our knowledge. For example, hormonal studies of these aminals would determine whether the cyclic behaviors described in the present study coincide with the ovulation period of the female. These studies are in progress by Nadler. The animals should also be studied in semi-natural habitats to determine the effects of such an environment on their behavior. Developmental studies would provide a better understanding of the sexual interactions observed in adults. These and additional studies should be carried out to determine an optimal captive environment for this species. In this way, the social relationships of established pairs could be improved to ensure maximum breeding success in captivity.

5
Birth and Parental Behavior

> The orang baby is just as helpless and dependent on his mother's protection as the human baby. On her milk for substitute for food, on her body for warmth. The orang baby's reactions—screaming for food and whining in case of discomfort, affection towards mother, slow orientation of vision and movement—are the same as those of a human baby. The only fundamental difference is the ability and need to grip, both with fists and feet, to the mother's body and hair. The baby becomes helpless without its mother also, because his limbs have lost their natural hold. It is essential to provide a substitute to hold on to, because it is through all four limbs that an incentive is given to move and climb, to learn and develop.
>
> (Harrisson, 1962, p. 93-94)

In this chapter I have reviewed several reports of birth and early development, in addition to the information which derives from our own research efforts. Much of the published literature is imprecise and subjective, but it is possible to put together a reasonably accurate composite with some effort. Moreover, there are some field data on parental behavior which can be compared to captive findings. In several instances, I have quoted extensively in order to provide greater detail. I consider this material the most important for those who manage captive orangs, since it is the quality of the early social environment which doubtless determines the reproductive success of apes. Throughout this chapter I have sought to reinforce my views with the findings and beliefs of others. Therefore, the material in this chapter represents what I believe to be the most informed opinion to date.

PREGNANCY AND BIRTH

An early account of an orang-utan birth was reported by Fox (1929). The birth actually took place in 1928 when the Sumatran female *Maggie* gave birth to a male at the Philadelphia Zoo, the first documented captive birth of this species in the United States. Fox did not actually observe the birth but discussed the pregnancy, delivery and early suckling time of *Maggie* by collating the notes of the keepers. I quote from this report as follows:

About the middle of February, 1928, "Maggie" suddenly developed a severe constipation for which there was no apparent cause. Simultaneously all appetite left her. She had had an attack resembling bronchitis or perhaps pneumonia around the first of the month. This lasted about a week, but there seemed to be a complete recovery. Just prior to the time the constipation and loss of appetite were noticed, she seemed as bright and active as usual. Such laxatives as are usually effective among the monkeys—milk of magnesia, castor oil, Carter's tea, prune juice—were offered, but because she was not eating and was drinking very little she could not be induced to take them. During the next three weeks she ate a little better, but frequently would go the entire day with no food at all, and drank little or no water. She would go as long as two weeks without any stool, but usually passed a small, dry, constipated piece of feces about the size of a nutmeg about once in ten days. During this time her activities and general demeanor remained unchanged, but the keeper believed her arms and legs showed some loss of flesh.

"Maggie" had shared her cage with "Chief," the big male orang, since October 6, 1927. Pregnancy was suspected, and because of this, and because the keeper advised against using force, no attempt was made to catch and restrain her for the purpose of giving an enema. The first observation that aroused the suspicion of pregnancy was swelling of the nipples and general mammary region during January, 1928. The nipples were frequently noticed to be enlarged and firmer, usually in the early morning. At times it seemed that the abdomen was enlarged, but the natural potbellied contour of the orang precluded any positive decision on this symptom. After the acute cold in early February there was a distinct ridge or shelf of the enlarged abdomen just below the ribs, such as might be expected in a squatting animal with an enlarged uterus. She was noticeably thinner by this time. About the middle of April the definite ridge or shelflike contour of the upper abdomen disappeared, suggesting the settling of an enlarged uterus into the pelvis. All during this period there was perceptible increase in the width of the abdomen. Enlargement of the breast areas was less from about the middle of April until June. Early in March some slight swelling in the vulvar region was noted, and this remained until during July, when it began to increase steadily. Enlargement of the mammary region was also much more conspicuous by this time.

During July "Maggie" began to show a gain of flesh all over, with a marked increase in the size of the abdomen, particularly well up toward the thoracic region, with a tendency to "pointing" in the midline just below the breasts. Her appetite was excellent, stools normal, and her general behaviour all that could be desired in a normal, healthy orang. This continued through August and into September, but about the middle of September she became peevish and quite irritable.

When the keeper came on duty and made his first rounds on the morning of September 25, he found "Maggie" tenderly licking an apparently normal baby, still wet, and with cord and placenta still attached. Observations were limited as it was thought best not to risk disturbing her. During the day the little one was seen to move about in the mother's arms, at times reaching up toward her face. "Maggie" remained lying down the entire day, cuddling her baby close to her face, but making no attempt to get it near her breast. At times, she evinced a mild curiosity toward the placenta, picking at it and turning it over, but making no attempt to separate it from the infant. She took her usual amount of fruit, but would not touch water or any other liquid, although she was accustomed to taking one quart of milk a day. The following morning the baby was noticed as apparently searching for something with its mouth, at one time sucking on the mother's ear. The mother by this time, 6:50 a.m., on the twenty-sixth, was beginning to show more interest in the placenta and cord, picking up the cord with her lips, then removing it from her mouth again. At 10:00 a.m., estimated as twenty-eight hours after the birth, she took hold of the cord with her teeth, close to the baby's abdomen, severed it, and then pushed the placenta over against the bars of the cage. Immediately she took up the baby and held it against her breast. It promptly found the nipple and proceeded to nurse. A few minutes later, "Maggie" emerged for the first time from the shelter to which she had retired for her accouchement, and with the baby clinging to her breast, climbed to the roof of the shelter and began a thorough examination of her offspring. A rather curious incident, and one worthy of note, occurred at this time. The keeper, taking advantage of her absence from the den, gave it a thorough cleaning and put in a fresh supply of bedding. "Maggie" had been accustomed to sleeping in the west end of the enclosure, and it was there that the baby was born; but she saw the clean bedding forthcoming, she immediately gathered it up, took it to the *east* end, made up her clean bed, to which she and the baby retired for a midday nap. Apparently "Maggie" appreciated the value of aseptic measures.

October 2, 1928. Since the last note "Maggie" has been doing very well. She is evidently a very good mother. The baby is not vigorous, but can and does climb all over the mother and nurses frequently. As early as the second day, the baby was able to cling to the mother's hair without the aid of the maternal arms. It is not practical to ascertain the amount of milk, but since the baby nurses and the nipples stand out straight, it is apparently adequate.

The baby has not been seen to void, but on the fifth day, the first stool was noted; it was firm and yellow. The mother did not have a stool for several days after delivery. On the evening of the fifth day, she drank a pint of warm milk to which four tablespoonfuls of milk magnesia were added. On the two succeeding days, three tablespoons of milk of magnesia were added to the usual pint of milk. Early on the eighth day, a moderate amount of feces was discharged in balls.

It was impossible to ascertain the lochia.* On the fourth day, the mother left the den and bed mentioned above, and made a bed in the outer and larger exposed cage, where she has lain or sat with the baby a good deal of the time since. She was apparently not disturbed by visitors, either the garden personnel or strangers. Mr. McCrossen thinks her disposition is better since the delivery than before.

The placenta was roughly discoid, 17 cm. in diameter and a few mm. to 2.5 cm. thick. The cord obtained measured 60 cm. The whole mass was dried when obtained. was soaked in water a short time and then fixed in formalin. There were some tears in the body of the placenta as if the animal had chewed at it, but there was no considerable defect as if she had eaten part of it.

* A vaginal discharge, at first bloody then serous.

Table 5-1a. Menstruation and birth data of orang female *Beatle* at Berlin Zoo. (Adapted from Lippert, 1974)

Date	Event	Difference
Oct. 14, 1967	Menstruation	
Nov. 18, 1967	Menstruation	35 days difference
Dec. 14, 1967	Menstruation	26 days difference
Jan. 11–13, 1968	Menstruation	28 days difference
Feb. 17, 1968	Labia majora swollen	37 days difference
Oct. 14, 1968	Birth of female *Ramona*	
March 19, 1970	*Ramona* weaned	
June 1, 1970	Menstruation	72 days difference
June 29, 1970	Menstruation	28 days difference
June 31, 1970	Menstruation	32 days difference
Aug. 27, 1970	Menstruation	27 days difference
Sept. 24, 1970	Menstruation	28 days difference
Nov. 30, 1970	Labia majora swollen	67 days difference
July 13, 1970*	Birth of female *Bata*	
May 10, 1970*	*Bata* weaned	
Jul. 17/18, 1972	Menstruation	68 days difference
Aug. 14, 1972	Labia majora swollen	28 days difference
Mar. 29, 1973	Birth of female *Biggy*	

To this early description we can add the more recent contribution of Lippert (1974) and others. For example, in 1963, van Bemmel depicted briefly two births, which were observed in the Rotterdam Zoo, in which the female in labor almost completed a headstand. Van Doorn (1964) also described a birth in which he suggested a gestation period of 260–275 days. In 1970 Ullrich reported on orang-utans which "*assisted*" the birth of other orang-utans at the Dresden Zoo.**

Variation in normal birth behavior among orang-utans is considerable. Therefore data concerning normal and abnormal pregnancy, birth, and development in this species is important. Experiences and observations concerning the behavior and breeding of orang-utans in the Berlin Zoological Park were the subject of Lippert's paper, which I will discuss in some detail.

A first sign of conception, according to Lippert, is the mild swelling of the *labia majora*, which occurs in the second month of pregnancy as a reaction to the absence of the next menstruation. He further asserted that the genital appendages constantly enlarged during the course of pregnancy, relaxing immediately after birth. This

** This "assistance" is a highly subjective interpretation.

Table 5-1b. Menstruation and birth results of *Ulla*. (After Lippert, 1974)

Date	Event	Difference
June 22, 1965	Labia majora swollen	
Aug. 15, 1965	Miscarriage	
Oct. 2-4, 1965	Menstruation	48 days difference
Nov. 15–17, 1965	Menstruation	44 days difference
Aug. 4, 1966	Birth of male *Moro*	
Dec. 5, 1967	*Moro* weaned	
June 26, 1968	Menstruation	204 days difference
July 30, 1968	Labia majora swollen	34 days difference
Mar. 7, 1969	Birth of male *Vroni*	
Mar. 19, 1970	*Vroni* weaned	
May 6, 1970	Menstruation	48 days difference
June 15, 1970	Labia majora swollen	40 days difference
Jan. 27, 1971	Birth of male *Tamor*	
May 10, 1972	*Tamor* weaned	
July 25–26, 1972	Menstruation	72 days difference
Aug. 31, 1972	Labia majora swollen	37 days difference
Apr. 19, 1973	Birth of female *Dunja*	

**Bata* born July 13, 1971. Weaning date must be 1971 according to "days difference" between weaning and next menstruation.

"pregnancy-induced genital swelling" was described by Schultz in 1938 from his observations of a pregnant wild orang, which had been shot. It is not a *sexually aroused* swelling as in chimpanzees. This sign of pregnancy occurred 28 to 40 days after the last menstruation of the Berlin females, suddenly appearing in 24 to 48 hours. After three days it could be readily seen. From 1968 Lippert registered all signs of pregnancy in the Berlin females, especially labial swelling. Using these facts he found that some conceptions were considerably later than the last noted menstruation indicated, with the gestation period overrunning the 260–275 days as determined by van Doorn in 1964. For example *Beatle's* next menstruation in Berlin was observed 72 to 68 days after the weaning of her current offspring. With *Ulla* also there were few menstrual results to be evaluated because she was usually pregnant again after the first or second menstruation. For her the next menstruation took place 48, 204, 48 and 72 days after the weaning of her offspring. Lippert asserted that although the 204 days was a great discrepancy, no menstruation escaped them.

The acquisition of regular urine samples is extremely important in any breeding program. Martin *et al.* (1977) have emphasized the importance of making such a procedure regular and systematic, and

Table 5-2. Birth data from Berlin Zoo (after Lippert, 1974).

For Birth Nr.	Mother/Young	Dilations-phase Passage of mucous ball before the expulsion of the fetus	First labor pain before birth	Expulsion/ Birth of fetus	Expulsion of the afterbirth after the expulsion of the fetus
2	*Ulla*/male *Moro*	about 15 min.	not observed	Aug. 4-66—14.15 h	63 min. postnatal
3	*Toga*/female *Anka*	40.5 hours	not observed	Nov. 11-68—12.35 h forceps birth	immediately after birth
4	*Beatle*/female *Ramona*	59–62 hours	22 hours	Oct. 14-68—20.50-21.00 h	immediately after birth
5	*Ulla*/male *Vroni*	not noticed	—	Mar. 7-69, before 07.00 h	—
6	*Toga*/male *Teluk*	35–40 hours	—	Sept. 30-69—18.00 h	20 min. postnatal
7	*Ulla*/male *Tamor*	3.5–5 hours	15 min.	Jan. 27-71—10.40 h	15-20 min. postnatal
8	*Beatle*/female *Bata*	—	about 15 min. before	July 13-71—12.45 h after birth	immediately
9	*Beatle*/female *Biggy*	not noticed	—	Mar. 29-73 between 05.00 and 06.00 h	—
10	*Ulla*/female *Dunja*	not noticed	—	Mar. 19-73 between 05.00 and 06.00 h	—

early enough to establish hormonal baselines for individual animals in the collection. To illustrate this, Martin and his colleagues listed the following questions which can be answered when urine samples are available:

1. We have had a pair of adult gorillas for several years without any breeding success. Is there any defect in the male or female which may explain this? Can anything be done about it?
2. We have two young orangs nearing sexual maturity. Have they actually reached puberty?
3. We have just received a new male chimpanzee to put into a breeding group, but we do not wish to put any existing pregnancies at risk. Are any of the adult females already pregnant?
4. We have to move a female orang to a new cage. Is she already pregnant? If so, when is the best time to move her?
5. we have a male gorilla that has killed an infant in the past. When our female becomes pregnant again, we would like to separate the male about a month before birth. Can we obtain a reliable birth-date prediction?
6. We would like to ensure that our chimpanzee colony does not become too inbred. Can we obtain detailed advice on a permissable level of inbreeding and the best course of action to take?
7. Our female orang has been diagnosed as pregnant and is due to give birth in 4–8 weeks' time. The keeper has noticed bleeding from the vulva in the last few days. Is this a danger signal and, if so, what action should we take?
8. We have been offered a pair of orangs from another collection. Can we check whether they are in good reproductive condition?

The technique practiced by Martin is to acquire samples in a clean plastic syringe either directly from the cage floor or from a plastic insert in the floor drain. Although a larger sample is to be recommended, one ml is enough to make the appropriate tests, but the sample should be split into two portions for storage. Urine should be filtered through glass-fiber filter paper and immediately deep-frozen at – 20° C or below. According to Martin *et al.* (1977):

It is absolutely essential that urine samples should be collected regularly for any reliable diagnostic work. For examination of menstrual cycles, daily samples over *at least* two months are preferable. If samples are collected every two days, valuable information can still be obtained, but gaps of three days or more may prevent reliable interpretation. If the day of conception is to be determined accurately, daily or alternate daily samples must be collected until a firm diagnosis of pregnancy has been made. For monitoring of pregnancy itself, it is then possible to work with samples collected once a week or once a fortnight. For studies of puberty, sample collected once a month would be adequate, and for study of hormone levels in males a few occasional samples would probably be sufficient. In the latter case, the samples collected should be quite large (e.g., 100 ml. each), whereas all other samples can be quite small (p. 47).

Pregnancy in the orang-utan is often difficult to detect due to the additional fact that adult females are often obese in captivity. Moreover, menstrual bleeding is difficult to observe, and there is no *visible* genital swelling to indicate the onset or cessation of cycle changes. Therefore, with orangs, a reliable pregnancy test is required. Hobson (1975) recently evaluated four commercially available test kits for both gorillas and orangs. Hobson used a larger than recommended urine sample with the *Gravindex,** *Planotest Pregnosticon*, and *Pregnosticon* ** test kits. A fourth test, using the recommended urine level, was the *Sub-Human Primate Tube Test (SHPT).**** *Of these four tests, the SHPT* was found to be the method of choice for diagnosing pregnancy in apes. Hobson points out, however, that the *SHPT* method is the most sensitive of the four, increasing the risk of false positive results. These risks are probably worth taking since later tests and other indicators would confirm the earlier reading, permitting a more precise estimation of impending birth.

The date of an impending birth may be much more exactly determined in advance by noting changed behaviors as were noted in Berlin. In the first months of pregnancy, females were found to be rather unpredictable and sometimes aggressive. In the last three months, they were quieter. They lay down a lot, but changed their resting place often and appeared to be much "occupied with themselves".

Lipperts' staff was also able to determine pregnancy difficulties in the first three months with both *Beatle* and *Toga*. Both exhibited little appetite and vomited frequently. In *Beatle's* second pregnancy the caretaker had to force her to eat and drink. They were lucky that she remained sociable during pregnancy, and they were able to prevent a great loss of weight. In spite of this, this female lost 4 kg in the first three months of pregnancy, as she took only little food despite their efforts.† In the following months of pregnancy she ate better and

* *Gravindex:* Ortho Diagnostics, Division of Ortho Pharmaceutical Ltd., Saunderton, Bucks, U.K.

** *Pregnosticon:* Organon Laboratories Ltd., Crown House, Morden, Surrey, U.K.

*** *Sub-Human Primate Tube Test:* Ortho Diagnostics Inc., Raritan, New Jersey, U.S.A.

† Recall here the feeding and elimination problems mentioned by Fox.

gained weight. In the last weeks before birth she was again without appetite. However, when she was fed then, she took the nourishment considerably more willingly. This fact led Lippert to suggest that she let herself be fed because the keeper kept her company. With the female, *Ulla*, changes in eating habits were not observed.

BIRTH BEHAVIOR

Orang behavior related to birth is, again, extremely variable. In view of the little material available, it is not yet appropriate to make great generalizations.

In spite of our ability to predict an impending birth date, it is nevertheless worthwhile to observe the pregnant animal continually. *Ulla's* first birth (the first in the Berlin Zoo's history) took place without complication. When intervention was required in subsequent situations, it became clear that norms were required from the onset of pregnancy to birth, in order to distinguish the normal from the abnormal.

The following record of *Anka's* birth is paraphrased as follows:

There had been no change in *Toga's* behavior up to January 9, 1968. Since mucous was observed in her vagina, the keeper led her into the sleep-box and so isolated her from the other orangs. As we checked again about 22.00 hours, *Toga* unexpectedly reached out of the sleep-box and attacked a person, who was known to her, and even after repeated requests did not leave the night shelter. In the morning of the following day *Toga* left the sleep-box only to take food. She was able to swing, and disappeared immediately inside. She had pains. A continuous observation of the birth was not possible under these circumstances, since she was only visible when she raised herself with a labor pain. According to outward signs, the birth did not seem to want to proceed. The fetus, which until the day before had been sitting high, had clearly sunk overnight. She had, as always, an adhesion of the anogenital region. Judging by her behavior, she must have already had labor pains nights also. During the day she rested diagonally across the back part of the sleep-box, which was exactly wide enough, so that the female fitted from anus to nape of neck between the walls. If she were to remain in this situation at night the fetus could not be forced out even with the strongest labor effort . We believed at the time, that the impeded fetus in *Toga's* first birth could be attributed to her remaining in the sleep-box. We would like to assume that female anthropoid apes, even with the first birth, would not behave "unbiologically" to such an extent, for even on that day, *Toga* had stood up with each labor pain, even when they were still light, remained for a long time standing on all four extremities in the box and in this position actively supported the labor action. On the morning of January 11, 1968, she lay around more and more apathetically in the cage, rose with difficulty and did not actively labor any more. It was decided to sedate the animal, and fortunately we were still able to intervene in time to get the fetus with forceps. We found out that the fetus had been stuck above with cording about his throat in the entrance to the pelvis of the female. The pelvic ring of the female was therefore not sufficiently wide or was altogether too narrow.

In order to avoid complications, we had intentionally isolated the male orangs before the birth and they were allowed in only as an exception, because we could thus handle the female in labor with less danger. After the descriptions by Ullrich in 1970, we also feared a dangerous precipitant birth, for our males too reacted to every birth (though to be sure entirely differently).* They were then very excited and aggressive against the keeper and the females, and *Hans* annoyed the female in labor when he was let in. At first we made use of this behavior with *Toga*, when on the morning of 11 Jan. '68 she would not leave the sleep-box, and because of that we let *Hans* in to her.

Hans immediately went to *Toga* and clicked his tongue very excitedly. It lasted only a few moments, and the female was challenged to come out. *Hans* at once made attempts to cover her, but *Toga* protected herself and again and again eluded him. Now it took a great deal of trouble to separate him from the female.

At the same time *Jessup* was similarly stimulated in the neighboring cage by the odor of *Toga's* birth sweats, so that he behaved aggressively towards all persons, who found themselves in the area of the grate, and traveled along the horizontal bars, imposingly and threateningly. Occasionally he sounded his call signals, obviously to alleviate his tension.

With *Ulla's* third birth, we were able to observe distinct labor pains, during the "dropping" of the fetus, which are sufficiently well known in the area of human medicine, but it was not clear to us until afterwards that they were really only "dropping" pains. In the beginning we believed it to be the beginning of the birth.

Often the birth of an orang is a surprise to all concerned and in these instances the indications of impending birth are usually a refusal to eat or drink, lethargy, and the passage of mucus from the vagina. Upon the onset of labor the passage of the infant is generally facilitated by the action of the mother. Lippert notes as follows in the 1972 birth of *Tamor* to *Ulla*:

When the fetal sac was out approximately 10-20 cm in size, she set to work with hand and opened up with fingers and the head of the child emerged. With the next pain the young animal was there, she let it slide into her hand and held it up immediately, sucked it clean on face. The young one cried immediately when it was born. *Ulla* held the young one in front of her face, clasped her lips over his face and sucked the embryonic fluid away. With the last pain *Ulla* grasped the young one with her hand and presumably pulled it and in this way supported the labor action.

---

* This view is contrary to our experience and will be discussed in detail.

The propensity of females to lick the face of neonates (that being the first bodily part which is visible) has often been interpreted as altruistic artificial respiration on the part of the mother, but this interpretation cannot be uncritically accepted.

Such behavior as described here demonstrates that even with one and the same female, each birth can be quite different.

In addition to these births, a difficult and a forceps birth were also described in Lippert's paper. Once even a labor caused by the dropping down of the fetus was noted. It was striking that *Ulla* behaved quite differently during this dropping labor than in the expulsion labor in the birth phase. She appeared "nervous", scratched herself again and again on her stomach, cramped up her hands during one pain, grimaced strangely and "whimpered in pipe-like tone" now and again. *Ulla* supported the dropping labor, in that she pressed thighs and legs with her hands against her body or pushed her stomach against the top of the table. During the birth, on the other hand, Lippert heard no sound from her, the extremities were loosely held and she inspected the vulva again and again with her finger. Also it appeared that in such an expulsion the entire sensorium was oriented "toward the back". By way of comparison, Lippert observed several times that, without outwardly visible reason, *Ulla* swung wildly through the cage between the occurence of two pains, which occurred during the dropping down of the fetus.

These vigorous movements may have been efforts to calm herself, bringing about the changes in the fetus's position as it dropped down. However, it is also possible that such movement was also an expression of pain. As Lippert noted, *Ulla* apparently found the labor associated with the dropping of the fetus very unpleasant and therefore retreated to the sleep-box, which was closed. Her wild swinging may have been to counteract the unpleasantness. Lippert further contended that *Toga* tried to calm herself during a rest in labor by completing a series of somersaults. Ullrich (1970) also described such behavior in a female orang.

Naaktgeboren (1971) suggested that the most important rule to follow in an anticipated birth is to prevent nervousness and preserve peace at all times. Lippert argued that excitement leads to activity, and dominance of the sympathetic nervous system which may *inhibit* labor. In contrast, Ullrich (1970) suggested that activity (induced, through stimulation by a male) *hastened* the birth process. These two views are clearly in opposition.

The most pleasant births are of course the uncomplicated ones, as represented by *Ulla* and *Beatle's* second birth at Berlin. During the expulsion of the fetus these females lay on their backs, their legs separated and let the fetuses glide into their hand or even actively pulled them out. The females appeared to learn with each birth to behave appropriately during labor. As Lippert cautiously expressed it, "with *Ulla's* third and *Beatle's* second, there was already a certain routine." In the Berlin group, *Toga* had especially difficult births. With her second birth she exhibited a whole spectrum of activity from headstand to lying on back. With the first pains she pushed lying on her back, the vulva pressed tight to the side surfaces of the table, just as she had done several times during her first birth in the sleep-box. Elder and Yerkes (1936) similarly reported a chimpanzee which sat on her buttocks three times during labor. As I have previously stated. The alleged incident of assistance in the birth (Ullrich, 1970) by a male, from stimulation to aiding in the birth itself, is a highly subjective anecdote. Lippert was similarly skeptical arguing that in the wild the possibility of a large number of arboreal accidents in births is unlikely, as the "four-handedness" of orang-utans assures a safe birth even in trees.

Each individual birth differs in duration, and as Lippert noted, the times from dilation to birth range from 15 minutes to 60 hours. Determining the first labor signs is considerably more difficult. Even with especially intensive observation there can be errors of judgment.

In a case study by Asano (1967), the birth of a young orang was briefly described as it occurred in the Tama Zoo, Tokyo. Let us now examine the details of this birth. Conceived in September 1964, the mother began to lose her appetite in October. In February of 1965, *colostrum* began to drip from her mammary glands. The birth of the infant on 6 May 1965 resulted after 261 days. In this, her first birth, the mother *Gypsy* bit off the placenta and consumed about one-fifth of it. When she did not nurse the infant, the ape keeper entered the cage and put the infant onto the breast, succeeding in the initiation of suckling some 37 hours after birth.* For one month, five to six times daily, the keeper facilitated suckling, thereafter needing only to do so when the mother exhibited occasional signs of swelling and pain during suckling periods.

The rather long dependency of orangs as compared to monkeys was noticed by Alfred Russel Wallace** (1856) who wrote:

It was curious to observe the difference between these two. The *Mias* (orang) like a young baby lying on its back quite helpless, rolling lazily from side to side, stretching out its four hands into the air wishing to grasp something, but unable to guide its fingers to any particular object, and when dissatisfied opening wide its almost toothless mouth and expressing its wants by an infantive scream. The little monkey, on the other hand, in constant motion, running and jumping about wherever it pleased, examining everything with its fingers and seizing hold of the smallest objects with the greatest precision...There could not be a greater contrast, and the baby *Mias* looked more baby-like by the comparison.

In a semi-natural game preserve, a birth to the Bornean female *Joan* was recorded by de Silva (1972). Seven months prior to birth, de Silva noticed that she appeared sluggish and remained inactive on the forest floor, her mammae and abdomen were enlarged and her genitalia dry and pale pinkish in color. The birth notes are edited and reproduced from de Silva as follows:

At about 0830 hours on 12 May 1967, Joan was observed behaving in an abnormal manner. She looked uncomfortable and attempted to lie down in various positions. She was seen to be straining. Although the cage in which she slept during the night was immediately opened, she refused to leave it, and grasped the iron bars and sacks with her hands and feet. She even attempted to bite—something she had not done before. As she preferred to stay in the cage she was not disturbed, but the door was left open. At about 1130 hours, while she was under observation, a sticky, colourless discharge started to drip from her vagina. Progressively it became thicker and more profuse and contained blood tinges. (The first stage of labour probably occurred during the morning; the gripping of iron bars and sacks being an indication of the pain of the uterine contractions). At about 1135 hours a clear, transparent sac pro-

* Certainly, nursing is not so frequent when so much of the early days of life are spent sleeping. As Harrisson (1962) has written:

During the early months of life, baby orangs sleep regularly during day-time. Ossy used to have naps of about thirty minutes after two hours of activity. He lay down to sleep in his basket, sometimes on my lap or, especially if it was hot, simply on the wire floor of his cot. He lay on either side or back, hands and feet gripping the edge of his basket, a rope, or a wooden bar of his cot. His night sleep was uninterrupted from dusk to dawn. (p. 99)

** When the animal was several months old, Wallace was also accustomed to leaving the young orang to hang unassisted for fifteen minutes at a time.

Table 5-3. Orang births in Berlin Zoo since 1965. (After Lippert, 1974).

Birth No.	Date	Mother	No. of births for mother	Young animal	Last menstruation	Days after	Labia majora swollen	Gestation time in days
1	Aug. 15, '65	*Ulla*	Miscarriage after about 3 months pregnancy				Jun. 22, '65	
2	Aug. 4, '66	*Ulla*	1st young	Male—*Moro*	Nov. 15-17, '65			262
3	Jan. 1, '68	*Toga*	1st young	Female—*Anka*	not noticed			—
4	Oct. 14, '68	*Beatle*	1st young	Female—*Ramona*	Jan. 11-13, '68	37	Feb. 17, '68	276
5	Mar. 7, '69	*Ulla*	2nd young	Male—*Vroni*	Jun. 26-27, '68	34	Jul. 30, '68	254
5	Sept. 30, '69	*Toga*	2nd young	Male—*Teluk*	Dec. 6-7, '68	74	Feb. 18, '69	297 ?
7	Jan. 27, '71	*Ulla*	3rd young	Male—*Tamor*	May 6-7, '70	40	Jun. 15, '70	266
8	Jul. 13, '71	*Beatle*	2nd young	Female—*Bata*	Sept. 24-25, '70	67	Nov. 30, '70	292 ?
9	Jan. 29, '73	*Beatle*	3rd young	Female—*Biggy*	Jul. 17-18, '72	28	Aug. 14, '72	255
10	Mar. 19, '73	*Ulla*	4th young	Female—*Dunja*	Jul. 25-26, '72	37	Aug. 31, '72	268

truded from the vulva. When this appeared, Joan was in semistanding position, her knees were slightly bent and her legs were wide apart. As the sac emerged she held it in her hands but made no attempt to manipulate it. About four or five seconds later, the vulval aperture widened and the sac ruptured. Joan grasped the sides of the cage with her hands and adopted a squatting position and began to whimper and cry; tears were visible. The baby was then expelled, and its head appeared with its face upwards. As the delivery progressed it was noticed that the hands of the baby were flat against its sides. The whole process took about a minute, and as the baby came out Joan took it in her hands. It was covered with mucus and blood; its eyes were closed and its body pale. Immediately after the birth Joan licked its face clean and pressed its cheeks so that it opened its mouth; then, placing her mouth on that of the baby, she blew three or four times into the baby's mouth, but I believe that prior to this act a normal respiratory rhythm had already been established naturally). The baby's fingers and toes began to move and the mother commenced to lick its body. While this was being done the baby's eyes and mouth opened and shut several times. Joan then chewed the umbilical cord and severed it well away from the baby's umbilicus. Holding the baby with one hand, Joan started straining and pulling at the umbilical cord which, after some manipulation, was expelled together with the placenta. It was first sucked and then eaten without hesitation. By this time the baby was attempting to position itself upon the mother. It gripped her fur and went off to sleep. The baby cried soon after it was born.

I observed Joan and her baby four hours later. She carried the sleeping baby close to her breast and at times nibbled at the umbilical cord. At birth the baby was about 30 cm (12 in.) in length, but no attempt was made to weigh it. When bedding was placed in the cage Joan promptly covered herself and the baby with it, although before doing so she drank a bottle of milk fortified with protein additive. Under observation, she appeared to be frightened of strangers and upset; observations were therefore discontinued. At about 1500 hours the next day Joan, hugging the baby to her breast, left her cage and proceeded in the rain towards the edge of the forest. She stopped near the bole of a tree and before climbing it, transferred the baby from her breast to her right shoulder, so that it would not get hurt whilst she climbed. She then covered the baby's head for a few seconds with the palm of her hand as if to ward off the rain. Then, steadying the baby on her shoulder with one hand, she climbed the tree, took refuge among the branches, remained there in the heavy rain, making no attempt to come down. The next day, at about 0600 hours, she returned to Sepilok holding the sleeping baby to her breast. It was observed that the remaining portion of the infant's umbilical cord was missing. Apparently the mother had chewed it off an inch or so from the umbilicus.

For more than six months after the baby was born Joan would not countenance any human or animal interference, but gradually she appeared to realize that the staff meant no harm to her infant. She would, if her confidence was gained, permit members of the staff to come near and stroke her offspring. On one occasion, the hand of a visitor was taken by Joan and placed on the baby's head. In 1968, Cocoa, a sub-adult female, played with the baby while Joan looked on. Joan has up to now (mid-1970) nursed and weaned her baby without any assistance.

Joan had been observed to be pregnant on 29 October 1966, and she delivered her offspring on 25 May 1967, an interval of 209 days. Assuming that the gestation period is 275 days, conception would have occurred about 24 August 1966.

An extremely interesting case of a multiple birth was recorded by Heinrichs and Dillingham in 1970. In this case, the primiparous female *Molly* exhibited several episodes of "grunting" two weeks prior to birth, interpreted as false labor. On the day of birth she took her normal food, resting on her belly, later grunting regularly at 5 to 10 minute intervals. The first fetus dropped at 1320 with the second arriving at 1420.

In this trio, the infants were not observed to nurse within 20 hours after birth, even though the mother in every other way cared for both infants continuously. At this point, the authors report that the infants were removed from the mother. Interestingly, when comparing this report to the experiences of others (cf. Swenson, personal communication) involved in orang-utan care, 20 hours without nursing is not cause for concern especially in a primiparous birth. * One cannot blame concerned medical authorities for caution, especially with a rare birth such as this one. However, this can be considered an example of a premature removal, an altogether too frequent event in zoos. Elsewhere (Maple, *et al.,* 1978) we have argued that a proper understanding of the variability and idiosyncrasies associated with maternal care would assist zoo authorities in making informed decisions about infant removal. Premature removal deprives the offspring of a normal rearing environment activating the syndrome which results in later sexual lethargy and/or maternal inadequacy. As Perry and Horseman (1972) have correctly pointed out:

A high proportion of young orang-utans are raised in isolation after they are taken from their mothers. Many are taken shortly after birth. Some zoo men are con-

---

* Occasionally, very young infants will have trouble suckling properly. Suckling can be elicited if the nipple is gently manipulated in the infant's mouth while holding its head firmly. The caretaker's hands should stroke the infant as the mother would in order to facilitate nursing and relaxation. As I have repeatedly emphasized, knowledge of the species-typical maternal repertoire will permit an appropriate simulation of mothering. The caretaker must not be impatient in this process.

Once the infant is nursing properly, surrogate maternal care can be provided by contact with the human caretaker and mechanical cloth surrogates as I have advocated in Chapter 7. To prevent digit-sucking, pacifiers can be utilized although care must be taken to insure that they are not swallowed. It is important not to prolong hand-rearing, and introduction to conspecifics should be achieved as early as possible. Weaning should not be premature, but introduction to solid food can be initiated at six months and accomplished by nine months.

cerned lest the hand-raising of zooborn orangutans further complicate the breeding problem. Many infants are handled like human babies, associating only with humans in the early months of life. When they are returned to the zoo, it is usually to a lonesome cage, since there are long odds against the zoo having a compatible cagemate. The managers of some collections advocate hand-raising and, indeed remove all infants as a matter of course, regardless of whether the mother seems capable of providing care. (p. 107)

One factor which must be taken into consideration in cases where infants are removed prematurely is the compelling presence of the nursery itself. Built from the contributions of a concerned public and staffed by experts trained in infant care, the nursery function is best fulfilled when there are plenty of infants to care for and to see. Baby apes are especially appreciated by a public which yearns to give them affection. Nurseries, though clearly necessary, ironically become self-fulfilling prophecies which, by their very existence, can compel well-meaning officials to prematurely remove infants from their mothers. Of course, in every instance of infant removal, a decision must be made. Whatever the outcome, the decision will be questioned by others. It is my hope that, by raising these issues and providing the information herein, such decisions will be made with objectivity and knowledge at hand.

THE SOCIAL NETWORK AND SOCIOSEXUAL DEVELOPMENT

As David Horr (1977) has pointed out, the developing nonhuman primate has three basic sources of "input" for socialization. The first of these are parental inputs which may differ depending on the respective roles of mother and father. These parental behaviors relate directly to the survival of the infant and the ultimate reproductive success of *both* parents. Thus the survival of the infant contributes directly to the number of genes the parents transmit to the next generation.

Secondly, inputs of *inclusive fitness* also contribute to infant development. By these terms we mean the socialization provided by close relatives other than parents; e.g., siblings, aunts, and uncles. Presumably, the genes held in common are ultimately responsible for the probability of these inputs.

Finally, the third source of inputs, according to Horr, are those which derive from all other conspecifics. In this sense, the larger

Plate 1. Facial expressions of an infant orang-utan temporarily removed from its mother at the Atlanta Zoological Park (T. Maple photos).

Plate 2. Dorso-ventral copulation posture at the Atlanta Zoological Park. Note that the hands of both male and female orangs are gripping cage furniture (T. Maple photo).

Plate 3. Ventro-ventral copulation at the Atlanta Zoological Park (T. Maple photo).

Plate 4. Adult female *Sungei* examining fingers of infant *Merah* at the Atlanta Zoological Park (T. Maple photo).

Plate 5. Orang-utan mother-infant pair in naturalistic habitat at the Kingdom's Three Wild Animal Park (T. Maple photo).

Plate 6. Ventro–ventral copulation of male *Bukit* and female *Sibu* at the Atlanta Zoological Park (T. Maple photo).

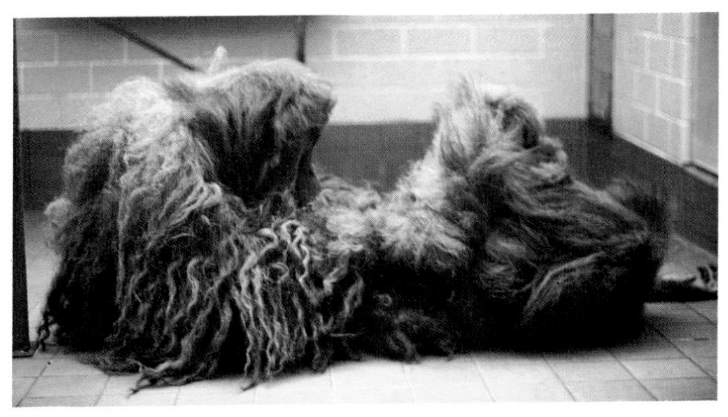

Plate 7. *Bukit* engaged in genital inspection of *Sibu* (T. Maple photo).

Plate 8. Copulatory thrusting by *Bukit*. Note the male's *pursed lips* expression (T. Maple photo).

social network contributes to the proper development of all animals born into the group. The total input system may be compared to what Harlow (cf. 1971) has called "affectional systems."

Table 5-4. Life stages of the Orang-utan according to MacKinnon (1974) and Rijksen (1978).

	Age		Wt. (Kg.)		
Life stage	MacK.	Rijk.	MacK.	Rijk.	Characteristics
Infant	0–2½	0–2½	2–15	2–6	Very small animals largely dependent on mother for food and travel (MacK.). Light pigmented zones around eyes and muzzle contrast with darker facial pigmentation; hair surrounding face long and standing out. Always carried by mother during travel; largely dependent on mother for food; sleeps in nest with mother (Rijk.).
Juvenile	2½–7	2½–5	10–30	6–15	Small animals independent in feeding and travel but still with mother (MacK.). Still mostly carried by mother, but makes short exploratory trips alone within her field of vision; plays, often alone or with peers; initially still sleeps with mother, but later it builds own next close to mother; towards the end of this stage mother may bear a new infant and attention for juvenile weakens. Facial characteristics as in infant (Rijk.).
Adolescent	7–10	5–8	20–40	15–30	Medium sized independent animals (MacK.). Hair surrounding face still long and standing out; initially facial coloration has the obvious light patches, but changes to completely dark; changes teeth; male and female are difficult to distinguish unless there is full view of ano-genital region. Frequent contact with mother; seeks contact with peers, plays with them and moves about with them in adolescent groups; cautious during contacts with adults, especially adult males; sometimes still travels with mother; starts showing sexual behaviour; females are sexually mature at about 7 years (Rijk.).

Table 5-4. Life stages of the Orang-utan according to MacKinnon (1974) and Rijksen (1978) (continued).

Life stage	Age MacK.	Age Rijk.	Wt. (Kg.) MacK.	Wt. (Kg.) Rijk.	Characteristics
Subadult male	10–15	8–13/15	40–55	30–50	Large animals with hard rims of undeveloped cheek flanges (MacK.). Facial coloration completely dark, with hard rims of undeveloped cheek flanges; beard starts to develop; hair surrounding face short, not standing out but flattened against the skull; testicles completely descended. This stage commenced with sexual maturity and continues until individual is socially mature; avoids contacts with adult males (Rijk.).
Adult female	8+	8+	35–50	30–50	Medium to large animals usually accompanied by young animals (MacK.). Old females may develop a beard, and are difficult to distinguish from subadult males, if not accompanied by offspring, nipples enlarged. Usually accompanied by offspring (Rijk.).
Adult male	15+	13/15+	45–100	50–90	Extremely large animals with cheek flanges, beards, throat pouches and long hair (MacK.). Extremely large animals; maximal development of secondary sexual characteristics; cheek callosities, beard, throat pouches and long hair. Sexually and socially mature, travels alone, moving cautiously; characteristic vocalization is loud calling: 'long call' (Rijk.).

Reproductive behavior, from conception to the birth process, is a subject of extreme importance for successful primate husbandry and conservation. Of equal importance is the successful development of the offspring. In the case of the anthropoid apes, the two systems of sexual and maternal behavior have been especially vulnerable to the effects of captivity. In the gorilla, Scollay *et al.* (1975) have estimated that 90% of the infants born in captivity have been separated from their mothers during the first or second day of life. This trend is also

Table 5-5. Stages in the life histories of orang-utans (after Rodman, 1973).

Female	Male
1. Constantly attached to mother, feeding only within her reach.[1]	1. (Same as female)........[1]
2. Forages independently of mother but carried by mother over long distances. Sleeps with mother.[2]	2. (Same as female)........[3]
3. Forages and travels independently but always follows mother. Sleeps near, but not with mother.[1]	3. (Same as female)........[1]
4. Spends several days and nights away from mother, but remains entirely within mother's range and frequently reunites with her.[1]	4. (Same as female)........[3]
5. Age of first reproduction. Daily travel completely independent of mother, range overlaps with mother's range, occasional encounters with mother.	5. Period of initial sexual maturity. Leaves mother's range and the ranges of any known resident males. Travels widely making random contact with females.[1]
	6. Competition for females with settled males and age-mates. Age of first reproduction.[3]
	7. Maintains contact with a few females and ranges over a fixed area which overlaps the ranges of those females.

[1]observed by Rodman; [2]observed by Horr; [3]hypothetical

apparent in the orang-utan for which Klos and Klos (1966; cited in Lippert, 1974) determined that, at most, one in ten infants was raised by the mother until it was independent. This situation is discomforting since a lack of conspecific mothering during infancy has been linked to deficiencies in both sexual and maternal behavior in adulthood (Harlow, 1971; Maple, 1977; Maple, 1979; Perry and Horseman, 1976).

The optimal rearing environment for an infant ape (or any primate) is composed of a mother, and where possible,the presence of other adults and peers.* As Harlow (1971) has pointed out, the affec-

* Of course, the degree of participation and presence of others is determined by the social organization of the species.

Figure 5-1. Newton Hartman photograph, as it appeared in Yerkes and Yerkes, 1929.

tional systems are sequential, with each stage dependent upon the adequate development of the preceding stage. It is therefore necessary to research the developmental sequence from birth onward in order to understand the structure, function, and effects of the rearing

Figure 5-2. Orang-utan *Sungei* with male offspring *Merah* at the Atlanta Zoological Park (T. Maple photo).

environment, and the changing roles of infants, mothers, and those conspecifics which compose the relevant social environment. The work of Harlow and his associates at the University of Wisconsin, and Goodall (cf. 1965) and her coworkers have provided a wealth of information on the development of rhesus monkeys and chimpanzees, respectively. However, little developmental information exists for many other species, with data especially scarce for the gorilla and the orang-utan. Where notes do exist, they generally are derived from home or nursery-reared animals.

In a report published by Maple et al (1978), we carefully followed the development of a Sumatran orang-utan (*Pongo pygmaeus abelii*) born 5-21-76 to *Sungei*, and *Lipis*. The infant's name was *Merah*.

From the day of *Merah's* birth we observed the animals for one to two hours five days per week, but observations briefly ceased for the first six months of *Merah's* life. During observation sessions, a single observer sat within six feet of the enclosure, recording behaviors sequentially as they occurred through the manual employment of our behavior codesystem.

Table 5-6. Maternal behaviors of orang-utan as recorded by Maple et al, 1978.

Cradling: cradling infant on the floor or in arms.
Holding infant: holding infant either on the floor or in the air.
Pushing: pushing or sliding infant ventrally or dorsally on the floor.
Ventral riding: infant carried on mother's ventrum.
Dorsal riding: infant carried on mother's back.
Clinging: infant clinging to mother.
Nursing: common usage.
Facilitate nursing: mother moving the infant to the nipple.
Deter nursing: mother moving the infant away from the nipple.
Retrieve: retrieve infant from another animal or physical structure.
Present infant: present infant to another animal.
Break contact: contact broken between mother and infant, but proximity maintained.
Standing while grasping: infant standing while grasping mother.
Standing alone: infant standing alone.
Locomote with contact: infant locomoting (crawling, walking) while in contact with mother.
Locomoting alone: common usage.
Digit sucking: sucking one's own fingers or toes, or those of another animal.
Inducing manipulation: mother placing infant's hands, etc., in contact with a bar or pipe.
Hanging alone: infant hanging from bars by himself.
Proximity: a distance of one body width or less from another animal.
Withdraw: move out of proximity.
Touch: common usage.
Groom other: common usage.
Groom self: common usage.
Hand extension: extending hand towards another animal.
Hand-hand contact: common usage.
Mouth-mouth contact: common usage.
Hand-genital contact: cmmon usage.
Mouth-genital contact: common usage.
Olfactory inspection: sniffing one's own hand after contact with another animal, or putting nose to another's body.
Sleeping: common usage.
Brachiate: hand-over-hand locomotion.
Walk quadrupedally: common usage.
Walk bipedally: common usage.
Mount: mounting another animal in a copulatory position; genital-genital contact established.
Sexual present: presenting ano-genital region to another animal.
Ventro-ventral copulation: copulation *more hominum*.
Dorso-ventral copulation: copulation *more canum*.
Yawn: common usage.
Baring teeth: open mouth with teeth visibly directed towards another animal.
Grimace: teeth showing, mouth slightly open, corners pulled back.
Kiss squeaks: vocalization made by the intake of air through extended lips (see MacKinnon, 1974).
Long call: deep, rumbling vocalization (see MacKinnon, 1974).
Grunt: common usage.
Funnel face: maximal pursing of the lips.
Chase other: common usage.

Table 5-6. Maternal behaviors of orang-utan as recorded by Maple et al, 1978 (continued).

Displace other: assuming the physical location that another animal had occupied.
Bit other: common usage.
Hit/grab: contact with another animal with either an open hand or fist.
Wrestle: wrestle with or roll another animal.
Mouth fighting: reciprocal, nonaggressive biting.
Self-feeding: common usage.
Take food: take food from another animal's mouth or hand.
Give food: give food to another animal.

MATERNAL CARE

Subsequent to the birth of *Merah*, maternal care was manifest in a number of ways. A frequently recurring behavior pattern was the mother's oral manipulation of the appendages of *Merah*; hands/fingers, feet/toes, and penis. We first observed oral manipulation of the penis during the fourth week, but it undoubtedly occurred earlier (cf. Hess, 1973). In another mother-infant pair, mouth-genital contact was first observed when the male infant was 11 days old (Davenport, unpublished data). Although the contact varied, oral-genital contact typically occurred in one of three ways: (1) with *Merah* cradled in her arms, *Sungei* leaned over and initiated penile contact; (2) she held *Merah* away from her body by his arms with his penis at the level of her face, and used her lips to manipulate his penis, or (3) she initiated contact by holding *Merah* down, his back on the floor. The duration of these interactions were usually brief, but did occur repeatedly in some instances. Oral contact frequently resulted in a visible erection of the infant's penis.

Sungei also repeatedly "mounted" her offspring, a behavior we observed during the second week of *Merah's* life. Although this interaction was not preceded by the same sequence of behaviors in all cases, the mother generally exhibited considerable interest in *Merah's* genital region prior to the initiation of mounting. She generally began by mouthing his penis, then placed him supine onto the floor and held his arms and legs in place with her hands and/or feet. She would then situate her genitalia above *Merah's* erect penis and make pelvic thrusting movements against his genitals (see plates 16 and 17). This behavior did not usually last for more than 10 seconds, but was often repeated several times, interrupted by retrieval to her ventrum and genital inspection. During active mounting, *Sungei* either looked down at *Merah* or around the room. The infant's response was typically to kick and wave his legs and arms, and vocalize. The infant's struggle to regain ventral contact may account for the short duration of any given mount, and the mother's frequent retrievals. In both oral-genital and mounting contact, we can determine no linear trend during the first six months, except to say that it is a common, perhaps daily occurrence, having been recorded on 31 of the 103 observation days.

The first observed instance of a complete break in contact between mother and infant occurred during the 18th week (cf. Figure 5-3; Month 4). This break was preceded by a sequence of behaviors that have been observed throughout *Merah's* development. While sitting on one of the bars above the floor, *Sungei* held *Merah's* hands in her hands, and raised him up over her head putting his hands in close proximity to the ceiling bars. *Merah* then proceeded to support him

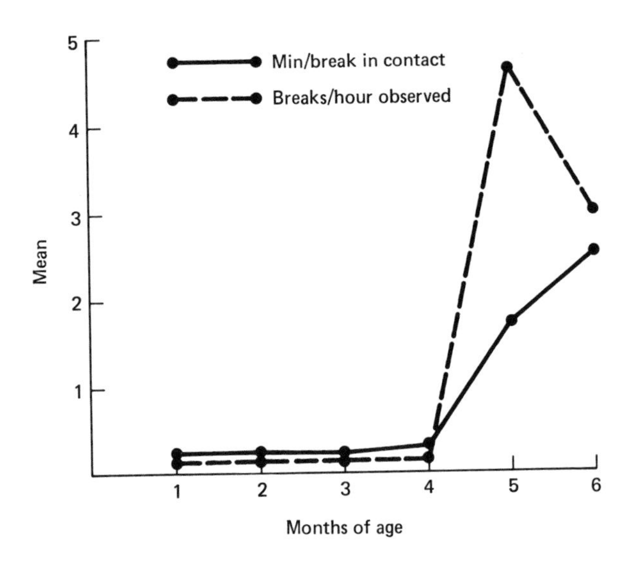

Figure 5-3. Monthly means of break in contact per hour observed, and length of time (in minutes) per break in contact between *Sungei* and *Merah*.

self by grasping a bar with one hand, holding his mother's hand with the other, and standing upon her shoulder. This entire sequence had been observed before, and was preceded in time by *Sungei's* early holding of *Merah* away from her body, and often up onto the bars (but with her hands over his). Hanging behavior was noted by Chaffee in an orang-utan mother at the Fresno Zoo when the infant was 10 to 12 months of age (Chaffee, 1967). It should be noted that at an early age, sometimes immediately after birth, orang mothers will place infants on their heads or shoulders and hold them there (orangs exhibit a tendency to put any novel object, and even familiar objects, onto their heads). It had been very difficult for us to resist the interpretation that she was "teaching" him to hang. Such an interpretation has also been suggested by other observers. Harrison reported the following from the Dusseldorf Zoo (Aulmann, 1932) referring to efforts by the orang mother to "educate" her baby towards climbing:

At the age of three months the baby does not move at all yet on his own, although the mother has started educating it towards climbing from its tenth day. She does this by taking the baby with one hand round the waist, and with the other, places its hands and feet round the bars of the cage. So far, the baby is very clumsy and does not grip well round anything except the fur of its mother, she also tries in another way to incite the baby to move on its own. She places it belly down, on the floor of the cage. Then, settling herself on a high shelf, she observes with great interest the baby's efforts to walk towards her—whining miserably while it does so. If the baby makes no progress, she comes down and gives it her finger to grip. Then she pulls it gently along the floor.

In discussing the mother's propensity to teach, Harrison further argued that:

It is quite wrong to assume that a young orang reacts to a large extent 'by instinct.' In my view the contrary is the case; most things are learned through direct teaching, and many fundamental attitudes—climbing trees for instance—are quickly forgotten by babies if they grow up on the ground to the extent of being frightened of any real height. Babies are taught first and foremost by their mothers, probably as long as four to five years; and, from the toddler stage (from about eighteen months old) they also learn from the incentive of their playmates. (p. 82)

Contact was broken on the first occasion when *Sungei* let go of *Merah*, sat a while and then moved away from him. The infant hung from the bars by his hands for 36 seconds before his mother retrieved him and returned to the floor. The average length of a break in contact markedly increased after the eighteenth week (see Figure 5-3). Other behaviors associated with early independence are represented in Figures 5-4 and 5-5 (*holding the infant in the air* and *inducing manipulation*).

Sungei did not actively share food with her infant, so far as we could determine, and he exhibited no interest in solid food until the 19th week of life. Before this time, *Merah* either sat passively in *Sungei's* arms, nursed, slept, or engaged in exploration, crawling around on his mother limbs. *Merah's* interest in food was demonstrated by his attempts to grab pieces which his mother dropped to the floor, or by his awkward attempts to grab food from *Sungei* while she was eating. Although the majority of these latter attempts were thwarted by *Sungei*, the infant occasionally succeeded in maintaining possession of a confiscated piece of food. When in his possession, *Merah* was observed to mouth solid food, but up to six months he was not observed to consume any.

Another maternal behavior which occured with some frequency was *sliding*, where *Sungei* placed *Merah* on the floor, either on his back or belly, and shoved him along the floor, pushing his arms with her hands.

OTHER MATERNAL-INFANT BEHAVIORS

Cradling

In times of inactivity, *Sungei* would typically hold *Merah*, cradled in her arms, against her ventral side if she was in a sitting position or, if in a reclining position, she would support his head with her arms while he lay or sat on the floor close to her body.

"Patting and Grooming"

Merah slept, nursed or looked at *Sungei* or around the room during grooming and patting sessions. Although grooming is a relatively infrequent orang behavior, and, when it occurs, is of short duration, it was during periods of inactivity that *Sungei* was observed to groom *Merah* (see also chapter 2), Harrisson (1962) observed grooming by a mother at the Dresden Zoo as follows:

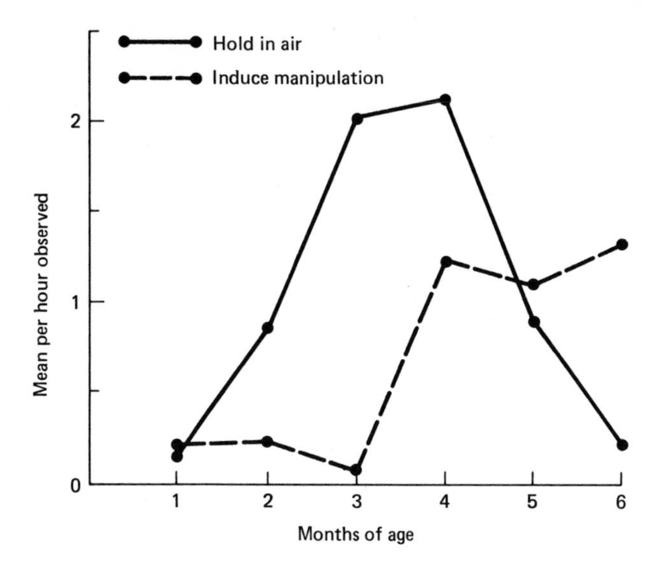

Figure 5-4. Monthly means of relative number of times *Sungei* held *Merah* in the air per hour observed, and the relative number of times *Sungei* induced manipulation of the environment by *Merah* per hour observed.

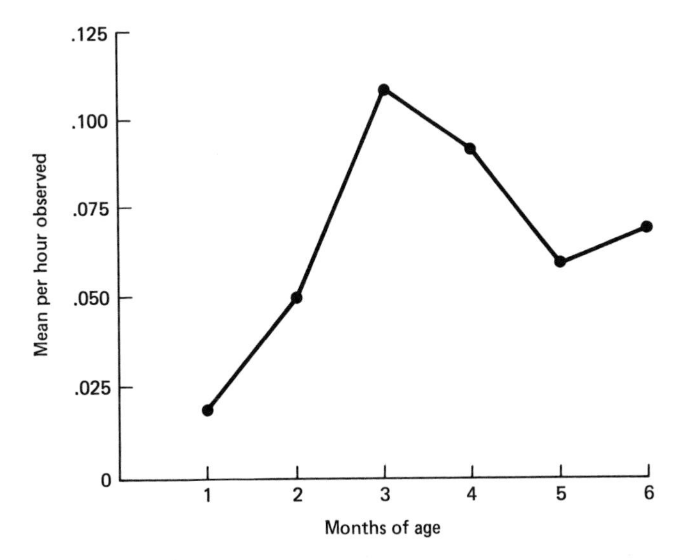

Figure 5-5. Monthly means of the proportion of time (in hours) *Lipis, Sungei,* and *Merah* were in proximity per hour observed.

. . . in Dresden Zoo . . . the mother, apart from keeping her baby's fur clean and "combed," also neatly bit off its finger- and toe-nails when they grew too long. The baby was then two months old and clinging to its mother the whole time. Later when the baby started climbing on its own, thereby sharpening its nails on branches, this became unnecessary. Harrisson, p. 99)

Activity While in Contact with Sungei

During or after feeding, *Sungei* would typically recline on the floor. If not nursing or sleeping, *Merah* climbed on and around his mother's outstretched limbs, or stood supporting himself while grasping *Sungei*. These behaviors were initially observed during the 15th week. At times *Sungei* would retrieve *Merah* and bring him to a cradling position on the floor from which he typically left to resume his previous activity.

Clinging, Carrying

During his first six months, *Merah* was only observed to ride on his mother's back twice as she moved across the room. The most frequent means by which *Sungei* carried *Merah* was with him clinging, unsupported by her, with both hands and feet in ventro-ventral contact.* Although there were times when the infant was observed to cling to *Sungei's* arm as she moved across the room, ventral riding was the most frequent mode of contact.

The infant mounting behavior of the mother *Sungei* is not idiosyncratic to her, nor to orang-utans. For example, we have observed the same mounting behavior in another 20-year-old female *Sibu* who frequently thrust against the erect penis of her 4-year-old offspring *Lunak* (Plate 16) in exactly the same way as did *Sungei* (Zucker, Brogdon, and Maple, 1976). The first author has himself witnessed this behavior in the female gorilla *Hazel*, mounting her two-year-old male offspring, at the Phoenix Zoo. Hess (1973) reported mounting of a female infant by the 24-year-old gorilla, *Achilla*:

* During a visit to the London Zoo in 1978, I observed an idiosyncratic carrying style by the experienced female *Kate*. While brachiating, she held her third infant suspended, its hand firmly clasped in her mouth.

During *Quarta's* first three years, *Achilla* repeatedly laid the infant on its back on the floor in front of her, placed her own genital region on that of the infant, then crouching down, either slid around with the child beneath her or performed rhythmic pelvic movements. (p. 520)

Hess noted that *Achilla* had been observed to mount two previous male infants in the same manner. At the Bristol Zoo in England, Hess also observed mounting of a six-month-old infant by its mother, *Delilah.*

Considerable oral-genital contact was also observed in these gorillas, often preceding the mounting of the infants. It seems clear, from Hess' observations of gorillas and our observations of orang-utans, that these great ape mothers are greatly interested in (and perhaps attracted by) the genitals of their infants. Moreover, there is some evidence that this interest is manifest differently according to the infant's sex. As Hess (1973) has written:

Goma, the mother of the male baby *Tamtam,* handled her infant's genitals in a far more differentiated manner than *Achilla,* the mother of the female baby *Quarta.* The difference is astonishing. It seems unlikely that the difference in maternal experience is responsible. The explanation may rather be found in the different sex of the infants. The erection of the penis and the wide arch described by the squirting urine in the male baby are reactions which are highly rewarding for *Goma* and which induce more intensive exploration. (p. 518)

While male genitalia may well be compelling stimuli to great ape mothers, it should be noted that the mounting behavior of *Achilla* was emitted toward *both* male and female offspring. Moreover, before we too rapidly assume the sex differences, it is necessary to consider our recent (and only) experience with an infant female orang-utan. *Hati* (Malay for "heart") was born at the Atlanta Zoo on Valentine's Day 1978. Within five hours after birth she was first mounted by her mother *Sungei* . Interestingly, she continued to mount *Hati* in the same fashion and with the same regularity as she had with the male *Merah.* Thus, our own limited experience with orangs does not support the generalization of a difference according to gender. Another interesting finding in the *Hati* study was that *Sungei* began to "hang her up" at the early age of one month, twice abandoning her and permitting her to fall to the ground. In both instances she quickly retrieved the infant, carefully inspecting its body. This was not a case of abuse, as *Sungei* is an experienced mother. I believe that this behavior can be best explained by referring to the psychological construct of "expectancies." *Sungei* had earlier lost *Merah* at nine months of age due to evidence of ill health (he later died). Her most recent maternal behavior was that of hanging the infant and moving away for long periods of time (30–45 minutes on occasion). In normal circumstances, where infants are spaced every three to five years, the most recent maternal behaviors of orang mothers involve less frequent contact with the young, and the offspring are the ones which control interaction. In my view, therefore, when an experienced mother gives birth in the presence of a relatively independent four to five year old she does not "expect" her newborn to behave like a six-month-old infant. The "expectancy" of *Sungei*, however, may well have provided the stimuli for *Hati* to be generalized to *Merah*. Thus, *Sungei's* "expectancy" may account for her aberrant behavior. To her credit, the mother quickly *learned* not to abandon *Hati*. Where any zoo should encounter a similar problem, I do not recommend removal of the infant unless it is clearly injured or not retrieved. Hay, sawdust, or some other soft floor covering might be a good precaution in the early stages of care, and is certainly a must if the infant has already been dropped. Here again, where the mother is alone, she may be more inclined to leave the infant since she does not need to protect it from interested competitors for the novel organism.

The behavior of our female orang-utan, *Sungei*, suggests to us that she derives some pleasure from mounting her infant. This may be a further example of the relative freedom from hormonal control which is the hallmark of anthropoids (cf. Beach, 1976; Nadler, 1976) and human beings. On the other hand, females that have resumed ovulation, may be expected to mount older offspring more readily when in estrus. However, we have no data to bear on this issue. Mounting of the infant may be compared to masturbation, a behavior which is reasonably common in captivity (Nissen, 1928; Maple, 1977). While the mother may enjoy mounting her infant, the act may also be beneficial to the normal sexual development of the young animal. The high degree of genital stimulation obtained from mothers is one element which is missing from the rearing environment of infants removed from their mothers. The lack of early genital stimulation may also contribute to the sexual lethargy which is too often characteristic of captive great apes.* In the orang-utans, early social deprivation appears to be especially troublesome since few viable offspring have been born to two captive-reared parents (Perry and Horseman 1976). Recently an orang-utan, *Pensi*, raised in the Atlanta Zoo group with her mother, father, sibling, other females and peers, successfully gave birth to an infant sired by a captive-born, mother-reared male. Born at the National Zoo, Washington, D.C., the mother properly cared for the infant for several days until its death from meningitis.

In a recent comparative study of a gorilla and orang mother-infant dyad at the Jersey Wildlife Preservation Trust, Kingsley (1977) logged 41.5 hours of observation time on each pair, distributed over three distinct sample periods as follows: 1.5–2.5 weeks, 5–6.5 weeks, 17–19 weeks. These data can be directly compared to our study of *Merah's* development during his first 6 months of life (Maple *et al.*, 1978).

Kingsley detected a steady decrease of ventral contact, whereby the predominate form of contact at 18 weeks was what she called "touch" contact. In this category, the infant sits or lies next to its mother while one individual rests a hand or foot on the other. By week 19, the infant orang-utan was in ventro-ventral contact less than 50% of the time, but the mother and infant were found to be *out of contact* less than 15% of the time. In the orang pair, 61.9% of the breaks in contact were attributed to the infant, but once separated, all movements away were initiated by the mother. In Kingsley's study, as in ours, the mother was observed to "hang" the infant onto the cage bars. Although little manual grooming of the infant was ob-

---

* Brambell (1975), on the contrary, has asserted that orang utans do not need to acquire reproductive skills:

> Unlike chimpanzees, orang-utans are not dependent on learning by example in order to mate in captivity. They just get on with the business as soon as they are old enough. Nor does the male restrict his attentions to the times when the female is at the fertile stage of her cycle. Copulation may go on right up to the time when the female gives birth. Thus the start of pregnancy cannot be judged by any change in copulatory behavior. (p. 241)

Regrettably, the only statement which I can agree to here is that some orangs copulate right up to the time of birth. However, it should be noted that it is the orang male which chooses to do so, not the female. Moreover, there is—as in so many other instances—great variability in the propensity to copulate with pregnant females.

served, the orang mother *Bali* did mouth her infants genitalia. As Kingsley indicated, the earliest *break in contact* in her pair was at 12 weeks, which is somewhat earlier than we reported for *Merah*. As Kingsley also rightfully suggests, early breaks in contact can signal the onset of breakdowns in the mother-infant relationship. However, many good mothers have been known to separate themselves from their infants early, but for brief periods only.

Detailed descriptions of the early rearing environment of great apes, under both natural and captive conditions, may be crucial to the captive propagation of these species. If we were able to specify the requirements for normal social development and adequate reproductive and maternal behavior in adulthood, it should be possible to devise methods to replicate maternal care in those situations where the infant cannot be reared by its mother. Furthermore, it may be possible to prevent premature separation of mother and infant, when the full range of maternal idiosyncasies are understood. This may be particularly important in the case of the orang-utan and the gorilla, about which little data have yet been published.

Regarding the effects of captivity on orangs, Jones (1977) has suggested that we look to the facts before assuming that captivity must necessarily adversely affect reproduction. He points out that the first birth of an orang in which *one* parent was captive-born took place at the Philadelphia Zoo in 1937 with the union of *Guas* and his own female offspring *Ivy*. Their subsequent young did not survive, but it is interesting to note that *Ivy* was mother-reared. Jones suggested that hand-rearing might have saved *Ivy's* infants, but it is likely that the propensity to permit the mother-rearing at Philadelphia contributed to *Ivy's* later success.

Jones also reported a successful second-generation birth rearing to two captive born parents at Rotterdam. No details are given regarding the rearing experience of the parents. Other instances of captive-born breeding successes are contained within the pages of the *Orang-utan Studbook* as maintained by Jones at the San Diego Zoological Park. As Jones further argues:

> One of the reasons for such a low number of breedings in which both parents are captive-bred is the difficulty of exchanging such large apes, and the fact that in most cases they are doing well in the respective zoos.

Figure 5-6. Adult female *Sungei* places infant *Merah* onto floor (T. Maple photo).

Figure 5-7. *Sungei* mounts *Merah* and emits pelvic thrusting onto *Merah's* genitalia. (From Maple *et al.*, 1978.)

As I have repeatedly maintained, it is not the fact of captivity *per se* which influences the subsequent mating and rearing success of captive born orang-utans. In my opinion, it is a naturalistic rearing experience—rarely achieved in captivity but not impossible—that facilitates later sexual and adequate maternal behavior. Mother-reared and socialized infants are a good bet to successfully breed and rear offspring even in captivity. Moreover, with experience, orang-utans become *better* mothers in captivity and in the wild. A possible exception are those instances where the prolonged stress of an inadequate setting leads to abuse of an infant. We have documented such a case at Yerkes in Puleo *et al* (in press). It is therefore still the *quality* of the habitat which counts in captive breeding.

As Jones himself pointed out in 1977, there were at that time 182 proven breeders among the captive-living but wild-born orangs, while among captive-born specimens (an admittedly smaller population) only seven were proven breeders. This is, in my opinion further testimony to the positive effects of early mother-rearing.

To add to the confusion surrounding this issue, I was recently informed of a case at the Seattle Zoo in which two hand-reared twins, raised together, successfully mated in adulthood. As this book goes to press, the mother is successfully caring for the offspring. The zoo veterinarian, James Foster, attributes their success to peer experience, further enriched by early association with two young gorillas

Table 5-7. Sexual behaviors of three differentially reared orang utans. (From Zucker et al., 1977.)

	Masturbation	Inanimate thrusting	Animate thrusting	Penis sucking	Hand-genital contact
Lunak (L) (5 years)	●	●	● S, K, L	○	● K
Loklok (LL) (6 years)	●	●	○	● L, K	● L, K
Kanting (K) (7 years)	●	●	○	○	○

Code: ● observed
○ not observed
S Sibu; mother of Lunak

(cf. Freeman and Alcock, 1973). As Harlow (1971) has demonstrated, peer-rearing can compensate for a lack of mother-rearing in monkeys. It is interesting to note that this case also contradicts the notion that orangs which are raised together will not breed. Thus, the idea of a brother-sister incest taboo cannot be conclusively supported. Perhaps the presence of the gorillas provided enough novelty to counteract the "familiarity taboo." In any event, this case illustrates the complexity of the problem.

SEXUAL DEVELOPMENT OF WILD ORANGS

As Rijksen has observed, a wild Sumatran orang first experiences genital stimulation by the actions of its mother and sibling. Infants are frequently the objects of interest as mother and sib touch, sniff, lick and stroke the infant's genitals during resting periods. As Rijksen notes, "such behavior, directed at the infant's genitals, was observed much more frequently than grooming activity directed at other areas of the body." Interestingly, Rijksen noted that infant female orangs have a rather conspicuous and prominent clitoris which superficially resembles a small penis. While other apes (cf. Van-Lawick Goodall, 1968; Hess, 1973) also exhibit interest in infant genitalia, one might expect orangs to show less sex-specific differences than, for example, gorillas where infant males appear to be more compelling stimuli

Table 5-8. Variables affecting sexual development in some Yerkes orang-utans. (After Zucker et al., 1977.)

Subject	Oral-genital stimulation by mother	Pelvic thrusting (mounting) by mother	Observation of adult copulations (as infant)	Observation of other copulations
Lunak	√	√	√	√ i, j
Tukan	√			√ i
Jinjing	√	√		√ i
Merah	√	√	√	√ i
Loklok				√ j
Kanting				√ j

√ : subject has experience
i = during infancy
j = during juvenile stage

than females. In fact, as we have noted, the earliest record of an orang-utan mother mounting her infant was *Sungei* mounting her female *Hati* within five hours after birth. (Maple and Zucker, in prep.) While maternal mounting of an infant has not yet been reported in the wild, we cannot rule out its occurence without further study. As Van Lawick-Goodall (1970) has pointed out:

> If a primate shows behavior in captivity which has not been observed in the wild, this by no means implies that it does not occur in the wild. (p.208)

Genital self-manipulation either by actual stroking or by the use of objects or other animals occurs regularly in wild infants according to Rijksen, occurring in males around six to seven months and emerging in females at about two years of age. Moreover, Rijksen discovered that wild infants appear to be interested in the genitals of their mother and sibling.

About masturbation, Rijksen reported that:

> Female orang-utans might masturbate by rubbing their fingers, their foot or an object along their clitoris, or they might insert their hallux or objects into their vagina. One adolescent female (*Sin*) was observed to suck and wet the finger she used during her masturbation. Males usually rubbed their fist or relaxed foot along their erect penis. Although the use of objects in masturbation was by no means restricted to rehabilitants, they did so more often than their wild counterparts. For instance, the sub-adult male *Sibujong* often pushed a hole through a leaf with his finger, which he then used to move up and down his erect penis. Young male rehabilitants might also insert their erect penis into small holes and crevices in branches and tree trunks along the way and make thrusting movements (p. 262).

MacKinnon also noted in his 1974 publication that young captive orangs exhibited sexual behavior at an early age.

> Male infants have been observed to rub their erect penis on the mother's (human foster-mother's) back (Harrison, 1960). Juveniles masturbate by rubbing the erect penis through chicken wire and on other impersonal surfaces. Juvenile females also masturbate manually or use pieces of wood or bark. I observed juvenile males at Sepilok masturbate on human hands by gripping the subject round the wrist with the feet and thrusting the erect penis between his fingers. (pp. 55-56)

The female orang does not exhibit cyclic genital swelling as do chimpanzees. However, their menstrual cycle, like the other apes, is approximately 30 days in duration (cf. Blakeley, 1969, Chapter 3 this volume). In late adolescence, about six to eight years, irregular cycles from 32–64 days (cf. Asano, 1967) are not uncommon. Rijksen noted that with the onset of puberty one female became increasingly intolerant toward other female peers, especially so when she was being consorted by a male. In suggesting this, Rijksen qualified his remarks by pointing out that little had been written about intolerance between captive female orang-utans. To set the record straight, our experience with the numerous female residents of Atlanta's Zoo and Yerkes Primate Center has revealed a well replicated phenomenon of female antagonism in captivity.*

Rijksen further pointed out that the context in which masturbation occurred differed in males and females. While males usually masturbated during resting periods or at times of little social excitement, females masturbated mainly when social excitement was high. Homosexual behavior among females and males was observed in rehabilitant orang-utans by Rijksen, including ano-genital intromission achieved by young males. Similar behavior has also been observed in fully adult males at Yerkes. A particularly bizarre form of sexual behavior was noted in the captive male *Lipis* who repeatedly attempted to enter the female *Lada* through a surgical incision made to alleviate a throat-sac infection (Swenson, personal communication) while housed at Yerkes.

SOCIAL REHABILITATION OF ORANG-UTANS

Since orang-utans, and all great apes, are in danger of extinction, their ultimate survival may depend upon breeding in captivity. Adequate rearing and social contact are essential for maturation and acquisition of the reproductive skills necessary for the propagation of the species. However, in captivity it is not always possible for infants to remain with their biological mothers. Occasionally, infant orang-utans require separation from their mothers, and are placed in nurserys for hand-rearing.

Early studies conducted by Harlow and his colleagues at the Wisconsin Primate Laboratories have demonstrated the deleterious

* Given adequate space, however, there are also distinct advantages to maintaining groups rather than pairs.

effects of social deprivation on primates.* Interactions with maternal figures and peers are important elements in the process of social development. Rhesus monkeys and chimpanzees reared in conditions prohibiting contact with conspecifics fail to develop adequate repertoires of sexual behavior (Harlow and Harlow, 1971); Turner, Davenport & Rogers, 1969). Animals such as these tend to display more self-directed and stereotyped behaviors than those raised by their own mothers either in the wild or in captivity (Cross and Harlow, 1965; Davenport and Menzel, 1963; Harlow and Harlow, 1971; Suomi and Harlow, 1972; Turner, Davenport and Rogers, 1969). These compensatory behaviors have been considered to be infantile responses which would ordinarily be directed towards mothers (Mason & Green, 1962). The development of such behavior is related to the absence or insufficient amount of stimulation that the mother ordinarily provides her infant as she grasps, hugs, rocks, and carries it during the early months of its life. (Davenport and Menzel, 1963). Indeed, self-directed and stereotyped behaviors are most common during periods of high stress, tension, or high arousal (Berkson, Mason and Saxon, 1963; Cummins and Suomi, 1976), times when mother-reared animals would normally return to their mothers.

Primate mothers encourage exploration by their offspring, while providing a safe place to return, should exploring become stressful (Harlow and Harlow, 1971). Deprived monkeys and apes receive neither of these benefits. They are fearful of novel stimuli, and are reluctant to initiate contact or explore novel situations, often exhibiting repetitive behavior.

Several attempts have been made to rehabilitate restriction-reared monkeys and apes; some unsuccessful, others at least partially successful. For example, Berkson *et al.* varied the environments of chimpanzees raised without mothers and found that their stereotyped behavior decreased in environments that provided opportunities to manipulate objects. Mason (1960) observed fewer responses indicative of emotional disturbance when rhesus monkeys were allowed access to a familiar partner or unfamiliar age-peer than when they faced either an empty cage, a cage containing a rabbit, or a cage containing an adult conspecific. After pairing deprived adolescent, male

*This section is a revised version of Puleo, Zucker, and Maple, in press.

chimpanzees with a wild-born, multiparous, adult female for a one-year period, Turner *et al.* (1969) discovered little improvement in the males' social behaviors. Chamove (1978) reported that infants and other isolates inhibited play and elicited fear in nine-month-old rhesus monkeys raised in partial isolation, but interactions with socially sophisticated age-mates elicited the highest level of play and least amount of fear. Suomi (1973) reported a significant recovery of both social and non-social behavior patterns, and a significant decline of self-directed disturbance activities in isolate rhesus monkeys following two weeks of partial isolation with an inanimate surrogate, and eight weeks of being housed with a same-sexed peer. Suomi and Harlow (1972) also reported significant recoveries of social and non-social play and exploration, and decreases in self-directed behaviors in isolate rhesus monkeys after "therapy" with younger females. They concluded that the males' behavioral levels were indistinguishable from those of their therapists.

Suomi and Harlow (1972) maintained that the crucial factor for successful rehabilitation was the nature of the social agents employed. Because isolate primates are easily frightened, and their behavioral repertoires are dominated by self-directed activities, the social therapists utilized must be those that will not elicit fear or avoidance responses. They must initiate social contact, but without aggression (Suomi, 1973; Suomi and Harlow, 1972).

In our efforts to rehabilitate nursery-reared orang-utan infants, we have successfully met the requirements of a good socializing agent by utilizing the propensity of adult female orangs to act as foster-mothers. Adoptions of motherless young have been reported for many species throughout the primate order, both in the wild and in captivity. Successful adoptions of young by adult females have been documented for the squirrel monkey (Eveleigh and Hudson, 1973: Taub, Lehrner and Adams, 1977), rhesus monkey (Deets and Harlow, 1974; Hansen, 1967) and chimpanzee (Palthe and van Hooff, 1975, van Lawick-Goodall, 1968). We used an adoption procedure in order to return nursery-reared animals to conspecifics and foster normal development.

Three infant male orang-utans, *Bulan*, *Jiran*, and *Anak*, had been housed in the nursery at the Yerkes Regional Primate Research Center prior to our efforts to rehabilitate them. *Bulan* and *Anak* had been in the nursery since birth, and *Jiran* since the age of 10 months. At the time of their introduction *Bulan* was 6 months old and not yet weaned. He was introduced to *Sungei* who was lactating, having lost her own infant (*Merah*), removed five days previously due to ill health. *Jiran* had been reared with his biological mother for the first 10 months of his life, but was then separated from her for health reasons, and housed in the nursery. At 2½ years of age he was introduced to *Paddi*, the 18-year-old multiparous female which had been without an infant for 5 years. *Anak* spent two years of his life in the nursery before he was introduced to a group of three orang-utans: *Jowata*, a 20-year-old lactating female; *Kesa*, her 10-year-old, nulliparous daughter; and *Jinjing*, *Jowata's* two-year-old son.

It is interesting to note that since the time of these adoptions both *Jowata* and *Paddi* have become sort of "super mothers," raising as many as three orphaned offspring at a time.

Bulan, *Jiran* and *Anak* were all observed as focal individuals. All social behaviors emitted or received by the focal individual were recorded sequentially using our standard code system. *Bulan* was observed continuously for one session of approximately three hours. *Jiran* and *Anak* were observed three to five times per week, 45 minutes per session, over a six-month period.

Bulan was introduced to *Sungei* at the Atlanta Zoological Park where *Sungei* was confined to the holding area of the enclosure, away from her adult male consort (*Lipis*). *Bulan* was first placed in a transport box which was positioned in front of the holding cage door. The doors were then opened, permitting *Sungei* the opportunity to reach into the transport box and contact him. The introduction of *Jiran* and *Anak* took place at the Yerkes Primate Center. Each of these infants was placed in one compartment of a home cage while the resident animals were locked in the adjoining compartment. The doors were then opened, giving the foster-families access to the infants and vice versa.

RESULTS OF REHABILITATION

Bulan. *Sungei* readily accepted *Bulan*, picking him up, examining him, cradling him on her ventrum, and directing various other normal maternal behaviors towards him. These behaviors included holding him out, inspecting him, and attempting to initiate hanging and climbing behavior (cf. Maple, Wilson, Zucker and Wilson, 1978). *Bulan*, however, had developed a persistent self-clutching response* in the nursery, and because of it he was unable to cling to *Sungei*. He was also unable to find the female nipple on his own, although when she placed him directly on the nipple he nursed briefly. After approximately eight hours we decided to remove the infant, fearing that his inability to cling might cause him to fall and become injured. We nonetheless considered this attempted adoption a successful effort because of the female's willingness to accept and nurture the unfamiliar and unrelated infant.

Jiran. Our second introduction involved *Jiran* and *Paddi*. This dyad exhibited behaviors typical of mothers and offspring. They spent a great amount of time in contact, of which a large portion was devoted to play. They also shared food, slept huddled together, and contacted each other in times of stress. *Jiran* initiated contact with approximately equal frequency during each observation session, while *Paddi* was much more variable, often avoiding *Jiran's* attempts to initiate interaction. This indicates that the adult female was the controller of social interactions within the dyad.

Anak. The third subject, *Anak*, was introduced under slightly different conditions in that he was placed in an existing familial group consisting of the three animals: *Jowata*, *Kesa*, and *Jinjing*. *Anak* received maternal care from both adult females, but received more from *Kesa*. Both females shared food with him, groomed him, and helped him climb to areas of the enclosure that he was unable to reach on his own. *Anak* initiated significantly more contact with

---

* Harrisson also commented on the development of this type of behavior:

> When Ossy came to us he used to grip with his feet around his own wrists because there was nothing else to hold on to. When he was given ropes and branches as well as the side bars of his cot, he gradually lost that painful attitude. (1962, p. 94)

Moreover, Wallace observed in 1869 the development of self-clasping in his little captive orang-utan:

> For want of something else, it would often seize its own feet, and after a time it would constantly cross its arms and grasp with each hand the long hair that grew just below the opposite shoulder (p. 66).

This grasping propensity is strongly programmed in orang-utans, and they rarely are out of contact with some tactual fixed object such as a branch, pole, or fence. Hornaday, writing in 1885, recognized this characteristic in an orang six months of age.

> For three of four days he would not allow me to hold him in my arms unless I would let him grasp some firm object with at least one hand. The action plainly showed that he feared that I would play a trick on him by letting him fall. (p. 382)

It is not surprising, therefore, that isolated orangs compensate for a lack of confidence and contact in their environment by clasping onto themselves.

Kesa than with *Jowata* (see Table 5-9). Under stressful conditions, he often clung to her, and at night he slept in contact with her. *Anak* spent the greatest amount of time in contact with *Jinjing*, the two-year-old infant. This contact was predominately play, most of which was initiated by *Jinjing*. However, as the study progressed, *Anak* increased the rate at which he initiated the play bouts. *Anak* initiated the fewest number of contacts with *Jowata*, but *Jowata* initiated considerable contact with *Anak*, mostly play. In general *Anak* initiated fewer contacts than he received from the other animals.

Table 5-9 reveals the percent of time *Anak* spent in contact with each cagemate. The amount of time spent with *Kesa* was less variable over observation sessions, whereas the amount of time spent with *Jowata* and *Jinjing* was highly variable. This variation seems to be a reflection of the amount of play involved.

Since *Jinjing* had lived with these two females since birth, he served as a control with which to compare *Anak's* behavior (Table 5-10). The high mean amount of time *Jinjing* spent in contact with *Jowata* reflects his attachment to her. Most of this contact was maternally oriented. *Jinjing* clung to *Jowata* frequently and continued to nurse. *Kesa*, on the contrary, spent more time with *Anak* than with

Table 5-9. Contact initiation during rehabilitation.

	Jowata	Jinjing	Kesa
Anak to:	72	181	120
To Anak:	154	337	95

Table 5-10. Mean amount of contact per hour of observation during rehabilitation. *Jiran* acts as the control subject. (seconds/hour)

	Jowata	Kesa	Paddi
Anak	396	217	
Jinjing	556	186	
Jiran			1336

Figure 5-8. Adult female *Sibu* mounts four-year-old offspring *Lunak* at the Atlanta Zoological Park. (From Maple *et al.*, 1978.)

Jinjing. The finding that both infants spent more time with *Jowata* than with *Kesa* may reflect the fact that *Jowata* has had more maternal experience than *Kesa*, and is generally more playful.

Anak and *Jinjing* can also be compared to *Jiran*. *Jiran* spent a great amount of time in contact with *Paddi*, since there were no other animals in the enclosure. However, the data are consistent with the high means of female interactions for *Anak* and *Jinjing*.

The most striking effect of our rehabilitation studies was the decrease in *Anak's* stereotyped behavior (Figure 5-9). He developed in the nursery a selfclutching, rocking stereotypy which persisted in the social group, but decreased rapidly and dramatically. Figure 5-9 depicts the amount of time per session engaged in this activity. With time this behavior was replaced by a variety of normal behaviors such as social and solitary play, and contacting a conspecific in times of stress. At the conclusion of the study, *Anak's* behavior had become virtually indistinguishable from that of a mother-reared orang-utan.

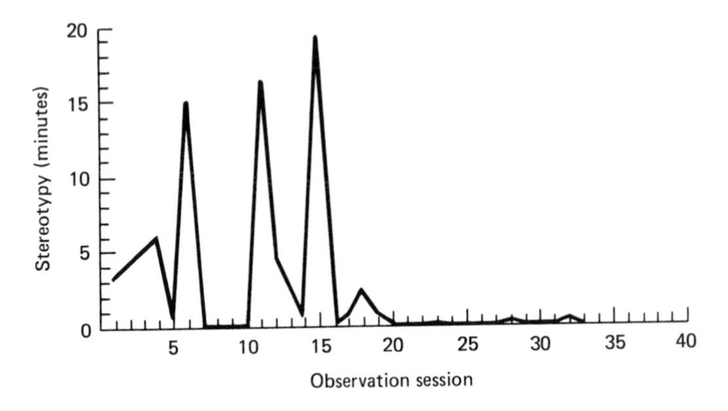

Figure 5-9. Stereotypy duration per observation session.

This research further emphasizes the importance of interactions with a mother or mother substitute. Both infants deprived of early maternal experience had developed pathological behavior in the nursery whereas the infant reared with its mother until age 10 months had not. While the most appropriate rearing strategy for young apes is to leave the infant with its biological mother, if she is clearly *not* capable of giving adequate physical and social stimulation, a social surrogate can be provided. Inanimate surrogates only provide limited physical stimulation, and are hardly adequate for infant socialization. Human surrogates may provide the infant with physical contact, nutrition, and some degree of social stimulation.* However, their efforts usually fall short of the quantity and quality of social stimulation provided by a conspecific. It seems, then, that the best method of rearing, other than rearing by the natural mother, is adoption by a conspecific adult female. Adoption provides stimulation for the animal essential for its development and socialization. Social stimulation also plays a major role in the decrement of pathological behavior. The adoption process may also give experience to a nulliparous female, useful practice for later motherhood.

---

* Here again, Harrisson (1962) suggests that peer-rearing is superior to that of human rearing:

It is so much better for young Orangs to grow up with a companion. They become more independent, get more fun. A companion—whether male or female—fills the gap that always remains between man and ape, however fond they may be of each other! (p. 91)

An intermediate rearing strategy was recently reported in the Brookfield Zoo's newsletter, (Brookfield Bison, 1973) as follows:

> . . .Brookfield Zoo is now hand-raising an infant orang-utan with a minimum of human contact. Hand-raised orangs frequently grow up socially and sexually maladjusted if, while young, they are exposed only to humans and never to other orangs. Consequently, the baby will be raised in front of the orangs enclosures in an ordinary baby crib . . . The crib has been placed in front of the enclosures of the other orangs to provide visual and vocal contact. Keepers feed the baby near the bars of the orang cages so she can see the other orangs at close range as well as smell and touch them.

An effort to avoid the problem of human imprinting (cf. Maple, 1977), this strategy deprives the developing infant of the consistent and varied physical stimulation that the mother provides. We have advocated (Maple, Zucker, Hoff and Wilson, 1978) elsewhere that human caretakers attempt to replicate the orang mother's activity as often as is possible, including regular carrying of the infant. At such times a fur vest may even be constructed and worn for "realism."

Referring once again to the orang-utan studbook, as compiled by Marvin Jones, of 42 captive-born orangs over 10 years of age in 1977, 20 have successfully bred and 22 have not. Of the successful breeders, only six have been males. Nine of the 22 unsuccessful breeders have been male. The vast majority of the births (41/49; ~84%) in which these animals took part were instances where the other parent was wild born. Thus, the rarity of two zoo-born animals producing offspring again becomes clear. As I have argued previously, the problem is not captivity *per se*, but the lack of early socialization provided by mother-rearing which prevents later breeding or subsequent maternal care.

The value and virtue of patience is best illustrated by the efforts of zoo veterinarian Howell Hood who intelligently intervened on behalf of the primiparous female Duchess at the Phoenix, Arizona, Zoo. Dr. Hood used operant conditioning techniques (food rewards and coaxing) to reintroduce her abandoned infant and encourage the appropriate maternal behaviors. As this book goes to press, mother and infant are together and doing well. Thus, by these heroic human efforts, another hand-reared orang-utan syndrome has been avoided.

PRESENCE OF THE MALE

The question I have been asked most frequently, by far, is "should the adult male be left with the pregnant female during birth?" It has been common zoo practice to separate females in the advanced stages of pregnancy in order to protect her from male interference, and to make human intervention easier if medical complications should arise. In formulating an answer for each question, I have always had to ask a few questions of my own. There is no simple rule of thumb to guide such an important decision. However, several issues must be carefully considered. First, a radical change in the physical or social setting contributes to stress,* and may adversely affect the birth or the subsequent care of the infant. Therefore, it is advisable to leave the female in her familiar surroundings for the duration of the pregnancy and birth. Moreover, where an adult pair has been compatible, it is worth a try to leave the pair intact. Our experience has been that adult males typically do not interfere with the birth. Newborns are generally ignored, although we have witnessed interest in the placenta, which males have been known to consume. Here, the caretakers may wish to be vigilant since it is possible that the two adults may struggle over this tissue. Certainly there is also the risk of sexual arousal,** although I would consider this to be a problem which is considerably less likely to occur.

The greater risks, it seems to me, are that an isolated, stressed female will abandon her young, or that upon re-introduction, the deprived male will injure the offspring in his mating frenzy. Clearly, there are many factors to be considered in the equation. If space is adequate, male interference is less likely, and female avoidance is possible. If separate but adjacent catch areas are available, the male can be present but restrained at the birth, then quickly reunited.

---

* Brambell (1975) has also made this point as follows:

Unlike mating behaviour, maternal behaviour does not come as easily to orang-utans. This may well be because the over-zealous manager tries to protect the baby by isolating the mother. This isolation may well be a cause of stress. It is relevant in this context to mention that London's first wholly successful mothering without any keeper help has been with a baby whose father was present at the birth. (p. 241)

** An interesting strategy for the prevention of male interference is suggested by events which transpired at the Seattle Zoo (J. Foster, personal communication). While the Seattle male was observing the birth of his first offspring, he became sexually aroused. An additional nonpregnant female was present also, and she became the object of a long and repeated copulation bout. Perhaps it is most useful to keep orangs in unsynchronized breeding trios.

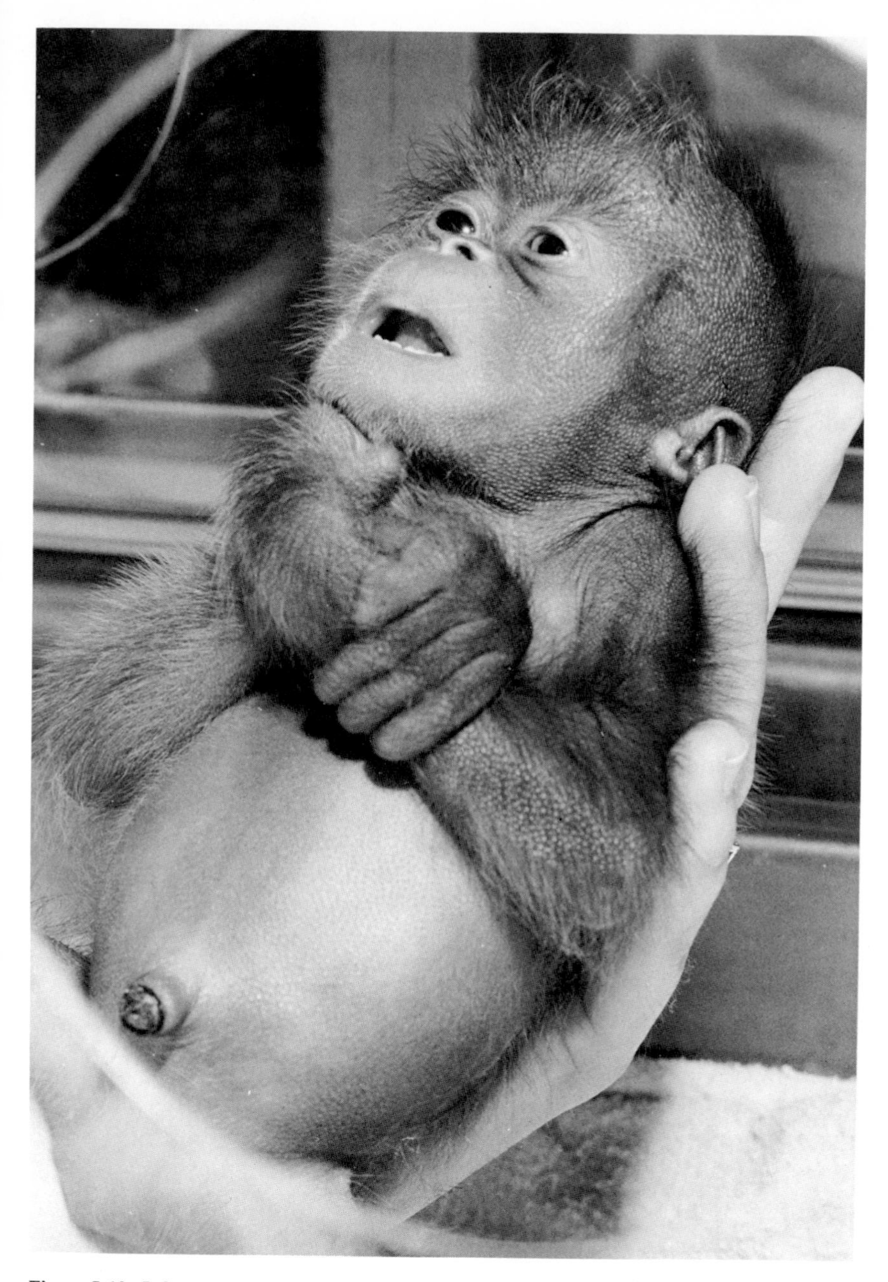

Figure 5-10. Infant orang-utan removed for hand-rearing at the San Diego Zoo. (San Diego Zoo photo)

Thus, it is the quality of the physical environment which permits the successful management of a birth. If one could anticipate the needs of the female and the staff, the ideal environment would include a visible but separate "birthing cavity" which only the smaller female could enter. The male could then see and be seen but not interfere. Within this area, video-tape cameras could permit caretaker vigilance without intrusion. To summarize, I do not generally fear interference from familiar adult males; my greater concern has been for interference by over-anxious human caretakers. There are times, however, where it is best to be cautious. Each setting presents different problems, and each institution relies on its past experience as a guide. Hopefully, as habitats are reconstructed to facilitate the birth and rearing process, we will breathe more easily with each successful reproduction.

HAND-REARING PROCEDURES

The first step when receiving an infant is to evaluate its physical condition. Records should be established concerning age, weight, sex, and medical problems such as visible wounds, dehydration, and respiration. Some authorities have advocated medication of the umbilicus, and trimming and tying it where necessary. Vitamins may be administered, and special attention should be given to prevent chilling. If an animal's body temperature is raised, many subsequent difficulties can be prevented. For orangs, hot water bottles in wrapped towels can be utilized, but direct contact with a warm body is especially valuable.

Once the animal's initial condition has been assessed and its problems ameliorated, daily feeding and contact comfort procedures must be implemented. The formula used is important, and in Table 5-11 I have compared the early diets which are followed at three institutions with considerable great ape experience. For the Philadelphia Zoo full details on their great ape diets can be found in Snyder (1976). Snyder (personal communication) has attributed the long life of the Philadelphia orangs *Guas* and *Guarina*, both of which lived in excess of 55 years, to dietary consideration. At about five years of age Philadelphia orangs begin to eat a diet composed primarily of monkey chow, but supplemented by smaller amounts of oranges,

Figure 5-11. Adult female *Sungei* grooming offspring *Merah* (T. Maple photo)

Table 5-11. Infant formulas prescribed for orang-utans at three different institutions.

Philadelphia Zoo	San Diego Zoo	Yerkes Primate Center
Infants, 0–6 months: Canned milk, multivitamins added. Feed every 3-4 hours. Last feeding midnight, until morning.	SMA* or Infamil *At 1½ months:* Gerbers strained cereal with apples and bananas, mixed in milk.	*FORMULA A and B*—SMA ½ and water ½ (birth to 12 weeks). Glucose w/ water first 24 hrs. 2½ oz. of Formula SMA per lb. of body weight per day. Birth - 4 weeks: 6 feedings per 24 hours as directed by schedule. 5 to 12 weeks: 5 feedings per 24 hours as directed by schedule.
6-12 months: Pablum 62% whole milk powder 35% wheat germ 2% vitamins 1% additions: cooked eggs raisins	*At 9 months:* canned milk, baby foods, vegetables, fruit	*FORMULA C* (12 weeks to 26 weeks)— Approximately 2-2½ oz. of Formula SMA per lb. of body weight per day. 0830: SMA ½ and water ½, pablum (to 7 tsp.), 1 tsp. egg yolk or 1 tsp. strained liver, plus 1 tsp. Vi-Dayin.

Table 5-11. Infant formulas prescribed for orang-utans at three different institutions (continued).

Philadelphia Zoo	San Diego Zoo	Yerkes Primate Center
oranges apples kale cabbage carrots		1300: SMA ½ and water ½. 1630: SMA ½ and water ½. Strained fruit or vegetables. 2300: SMA ½ and water ½. *FORMULA D* (26 weeks to 52 weeks). Same as Formula C except give more total food. Start whole fruits and vegetables according to changing needs of animal. *FORMULA E* (52 weeks to 78 weeks). 0900: 2 cups of Formula mixed as follows: (1 lb. whole dry milk per gal. water). To each A.M. cup add ¼ cup Pablum. To the *1st* cup on Mon., Wed., Fri., add 1 tsp. of Vi-Dayin and 1 tsp. egg yolk. On Tues., Sat., and Sun. add 1 tsp. strained liver to the 1st cup. 1630: 2 cups of above formulas, plus Jr. food, peas, carrots, beans, etc. in rotation. Whole fruits and vegetables according to changing needs of animal. 1200: ½ carrot and piece cabbage. 2 slices wheat bread. *FORMULA F.* Same as Formula E except more total food is given according to changing needs of the animal.

*Iron causes some orangs to suffer constipation during the first month of life. Infamil without iron may be preferred in these cases.

and other fruits. On this diet, browse is served five times per week. The type of browse varies, but it is considered to be especially important for the well-being of the animals. In Snyder's opinion, orangs must not be permitted to become obese when older, but need especially nutritious foods when very young.

Table 5-12. Top twenty world zoos in number of orang-utans born as of 1977 (adapted from Jones' *Orang-utan Studbook*).

Zoo	Number Born	Now Living	Have Since Bred
1. Yerkes Primate Center*, Georgia, USA	22.15	13.6	0.1
2. Colorado Springs, Colorado, USA	12.11	7.8	0.2
3. Frankfurt, West Germany	12.10	8.4	0.1
4. Rotterdam, Netherlands	10.6	3.3	3.1
5. St. Louis, Missouri, USA	10.4	2.2	0.1
6. London, Great Britain	8.7	3.4	0.1
7. Sydney, Australia	7.8	4.5	1.2
8. San Diego, California, USA	7.7	4.5	0.9
9. Calgary, Canada	7.2	6.1	0.0
10. Berlin Tierpark, East Germany	6.5	3.5	1.0
10. Berlin, West Germany	6.5	4.2	0.0
11. Duisberg, West Germany	5.6	4.5	0.0
11. Perth, Australia	5.6	3.5	0.0
12. Brownsville, Texas, USA	5.4	4.4	0.0
13. Dallas, Texas, USA	5.3	3.2	0.0
14. Amsterdam, Netherlands	5.2	4.0	0.0
15. Philadelphia, Pennsylvania, USA	4.8	1.5	1.2
16. Dresden, East Germany	4.5	3.4	0.0
17. Basel, Switzerland	4.4	2.2	0.1
18. Fresno, California, USA	4.3	4.2	2.1
19. Toledo, Ohio, USA	4.2	2.0	0.0
20. Memphis, Tennessee, USA	4.0	3.0	0.0

*Eight of these births have actually taken place while on loan at the Atlanta Zoological Park, which could therefore lay claim to a ranking of 13th among world zoos.

The San Diego authorities (Joanne Thomas, personal communication) agreed that nutritious foods early in life produced large and healthy orangs.* Both the Philadelphia and San Diego infants diets are administered as if the apes were human children. When in doubt, as the authorities agree, a standard pediatric diet for children can be used.

In summary, the proper diet is essential in producing healthy infants, preventing boredom, and in promoting longevity.

---

* However, well-fed baby apes may not be as desirable as some caretakers suggest. In a recent review of the human child-care literature, Lomax et al (1978) advanced the argument that well-fed babies often turn out to be obese adults. Since bottle-fed babies generally gain weight faster than breast-fed babies, this argument has been used in advocating a return to breast-feeding. By the same logic, the problem of obesity in apes may be compounded by hand-rearing in an animal nursery.

Where hand-rearing *must* occur, there are other important techniques which have been described. For example, Harrisson (1962) wrote:

> Fortunately, standards of infant care have improved considerably and it seems comparatively easy, these days, to bring up an orang by hand—although this type of upbringing is no ideal substitute for mother's milk and education. Two factors are important: that the infant naturally relies on mother's milk for a long period—probably at least four years; and that the mother introduces the infant to masticated fruits, chewing of leaves and branches, as well as to exercise at an early age. (p. 147)

Moreover, she suggested the following nine important steps in the human care of infant orang-utans:

1. To be off the ground.
2. To be able to grip and hold fast to ropes and branches, so as to exercise his limbs.
3. To have a constant fresh supply of leaves, suspended swaying over his head, to chew and play with.
4. To be out-of-doors.
5. To be out of the sun and out of the rain most of the day.
6. To have a blanket at night.
7. To be cuddled (but not too much!).*
8. To have no strangers—and above all no one with a trace of cold or lung troubles—breathing over it.
9. To have a regular routine of feeding times, bath and sleep. (p. 90)

There are many questions about orang-utan development which remain to be answered. For example, at what age do we begin to see

---

* In a telephone conversation, Joanne Thomas, who has hand raised apes at the San Diego Zoo for twelve years, also told me that apes should not be held too much lest they become spoiled. According to Thomas, by giving them time off "mother" their bodies also develop faster.

At the Philadelphia Zoo, following institutional hand-rearing procedures, surrogate "mother" contact was accomplished by the participation of round-the-clock caretakers, called "gorilla-girls" by Snyder. In any zoo, individuals such as these could be paid staff, students, or volunteers. Their interaction with the young apes provides a source of comfort and stimulation which is also entertaining to the public, where visual access is possible.

Caretakers have repeatedly asked me if it is necessary to replicate in minute detail every action of the orang-utan mother. The answer, of course, is *not* literally. By the use of a towel, the infant's body can be rubbed all over, simulating the action of the mother's body. Thus, genital stimulation can be accomplished without the *exact* duplication of effort.

rudiments of complex adult vocalization? Does the solitary, peripheral nature of adult males begin to develop in adolescence or even earlier? We also need to determine, once and for all, the *adequate* level of social experience necessary for the normal development of sexual and parental behaviors. Only by careful longitudinal research will these questions and others be finally and completely answered. Needless to say, we should not permit our lack of information to render us helpless. There is currently enough information to intelligently manage these apes in captivity and I will explore this in greater detail in Chapter 7.

6
Intellect of the Orang-utan

> The orang-utan which I saw walked always on two feet, even when carrying things of considerable weight. His air was melancholy, his gait grave, his movements measured, and very different from those of other apes . . . signs alone were sufficient to make the orang-utan act; but the baboon required a cudgel, and the other apes a whip, for none of them would obey without blows. I have seen this animal present his hand to conduct the people who came to see him, and walk as gravely along with them as if he formed a part of the company. I have seen him sit down at table, unfold his towel, wipe his lips, use a spoon or fork to carry the victuals to his mouth, pour his liquor into a glass, and make it touch that of the person who drank along with him . . . All these actions he performed without any instigation than the signs, or verbal orders of his master, and often of his own accord.
>
> Buffon, p. 169 of *The Naturalist*, 1830

Intellect is a very difficult concept to define. In fact, it was very difficult for me to select a term which adequately represented what this chapter is about. If we look to the authoritative Oxford Dictionary for assistance, we will discover the following primary definitions for the word "intellect:"

1. That faculty, or sum of faculties, of the mind or soul by which one knows and reasons (excluding sensation and sometimes imagination; distinguished from *feeling* and *will*); power of thought; understanding. Rarely in reference to lower animals.

2.a. An intellect embodied; a being possessing understanding; an 'intelligence,' a spirit.

In the pages which follow, I hope to provide sufficient detail to enable the reader to comprehend the learning potential of orang-utans in relation to their close relatives the chimpanzees and gorillas. Beyond the general definition provided above, we know that the intellect of an ape can be assessed in a variety of ways, suggesting that there are numerous dimensions to the construct. Although the published research does not permit me to enumerate all of the possible dimensions, several investigators have examined the learning abilities of the apes. I will separate these studies according to the concepts and operational categories employed by the respective experimenters.

The intelligence of the orang-utan has always been a matter of some dispute. It is the temperament of the species which likely accounts for these differences in opinion (cf. Chapter 3). Unlike the boisterous chimpanzee, the orang-utan presents a calm, calculating facade. According to Yerkes and Yerkes (1929):

> As a stage performer the orang-utan is rare by comparison with the chimpanzee. It is readily trained to manlike habits of eating, drinking and occasionally to care for its quarters, but it is not so well adapted either structurally or temperamentally to acts of skill and to entertaining performances as the chimpanzee. Authorities differ as to capacity for training, some asserting that it is even more easily trained than the chimpanzee, but the majority maintain the opposite . . . As a laborer or domestic servant the orang-utan is relatively successful. Apparently it takes more kindly to work or drudgery than does the chimpanzee, but seldom has it been so thoroughly domesticated and broken to manlike habits that it willingly and with good nature performs monotonous acts. (p. 134)

In reviewing the literature in reference to the orang, Yerkes and Yerkes further determined that with respect to overt activity it was commonly described as quiet, inactive, lethargic, sluggish, slothful, lazy, slow.* They further asserted that the orang gave the impression of a stolid, sad, depressed and phlegmatic attitude. The facial expressions of the orang, they stated, were indicative of a grave, sedate, serious, thoughtful, reflecting, brooding, pensive, and melancholy facade.

Despite these generalizations, which suffer from a lack of precision among other things, it is impossible to completely characterize the species, given the enormous temperament differences between age and sex classes as well as individuals. Young orang-utans, for example, are relatively active, energetic, playful compared to the generally sluggish character of older animals. Nevertheless, adults *can* be playful, and with proper motivation, opportunity, and experience, they

---

*A very interesting impression of the orang was generated in the book *Planet of the Apes:*
They bring the same characteristics to all these activities. Pompous, solemn, pedantic, devoid of originality and critical sense, intent on preserving tradition, blind and deaf to all innovation, they form the substratum of every academy. Endowed with a good memory, they learn an enormous amount by heart and from books. Then they themselves write other books, in which they repeat what they have read, thereby earning the respect of their fellow orangutans. (Pierre Boulle, p. 111)

engage in a life of considerable activity. As the Yerkes' (1929) so correctly asserted:

> As one observes for himself, and seeks to supplement his first hand information by searching the literature, he comes to suspect that adequate general description of the affective behavior of the orang-utan is a difficult and intricate task which up to the present has not been seriously undertaken by any adequately trained psychologist. (p. 151)

Thus, despite the many ambiguous and imprecise references to orang-utan temperament, the behavior of orang-utans during formal testing and in training sessions has not always helped to elevate its intellectual standing in the primate order. Suffice to say that the orang does not always perform with enthusiasm. Nonetheless, Sheak (1922) was led to remark:

> While the orang-utan is quiet and unobtrusive, and not as good an animal for exhibition purposes as the chimpanzee, I believe him to be almost, if not altogether, as intelligent. He is not always inventing countless new ways of amusing himself and working off a superabundant store of physical and mental energy, as does his African cousin, but when it comes to solving problems to satisfy his own needs or desires, and to doing things that are really worthwhile, he manifests wonderful intellectual power. (p. 50)

Clearly the principle problem to be overcome in discussing the intellectual abilities of any animal is to define at the outset precisely what criteria are to be used in the assessment. Unfortunately, when viewing the problem historically, the criteria are numerous and frequently unsatisfactory for comparison. Some representative categories will be reviewed here in the hopes that at least a few concrete hypotheses or generalizations can be made. We have in this chapter and in Chapter 3 already discussed the temperament of the orang, and this, together with physical limitations and species-typical motor propensities, puts contraints on the animal when it is tested. As in all efforts to measure the ability of an organism to learn, the appropriateness of the task and the motor/sensory adaptations of the animal must always be kept in mind.

A second opinion on the matter of orang intelligence was provided by Hornaday in his 1922 book *The Minds and Manners of Wild Animals*.

The orang is distinctly an animal of more serene temper and more philosophic mind than the chimpanzee. This has led some authors erroneously to pronounce the orang an animal of morose and sluggish disposition, and mentally inferior to the chimpanzee. After a close personal acquaintance with about forty captive orangs of various sizes, I am convinced that the facts do not warrant that conclusion. The orang-utans of the NY Zoological Park certainly have been as cheerful, as fond of exercise, and as fertile in droll performance as our chimpanzees. Even though the mind of the chimpanzee does act more quickly than that of its rival, and even though its movements are usually more rapid and more precise, the mind of the orang carries that animal precisely as far. (p. 73)

EARLY EXPERIMENTAL STUDIES

In their 1929 review of the early literature, the Yerkes mentioned the work of Haggerty (1913) as the first experimental inquiry regarding orang intelligence. Working in the New York Zoological Park, Haggarty gave two subadult females an opportunity to acquire food on a table out of reach. Given a stick with a hook on the end, both animals used the sticks to acquire the food. As Haggerty noted:

Had a human being exhibited this behavior for the first time, we should describe it most easily by saying that he perceived the relation between the food, himself, and the stick. (p. 153)

In comparing the orangs to the zoo's chimpanzees, Haggerty concluded that the former were superior in the learning tasks which Haggerty devised. The Yerkes evaluated Haggerty's work and concluded that it deserved "high credit and recognition as an initial experimental study of orang-outan adaptivity" (p. 183).

Furness (1916) also carried out some notable early research, demonstrating the imitative tendency, discriminative ability, perception, memory, and recognition ability of the orang. Furness was *wrong* in his finding that orangs could not draw certain figures, nor tie knots.

The most important early research on learning, however, was conducted by Robert Yerkes himself and published in 1916. In a study of "ideational behavior," Yerkes set out to learn how orangs solve novel problems, using the *method of multiple choices*. The problems were graded in difficulty, and dealt with relations among objects such as: (1) secondness from one end of the group of "reaction-mechanisms"; (2) middleness; and (3) alternation of end mechanisms. The orang apparatus consisted of nine boxes which could be opened and entered. An orang named *Julius* was used in this study and first presented with the problem of learning to select the box at his extreme left. Contrary to expectations, *Julius* persisted in choosing the *nearest* box. The solution to the problem came suddenly and, as Yerkes put it, *ideationally*.

> The curve of learning plotted from the daily wrong choices . . . had it been obtained with a human subject, would undoubtedly be described as an ideational, and possibly even as a rational curve; for its sudden drop from near the maximum to the base line strongly suggests, if it does not actually prove insight.
>
> Never before has a curve of learning like this been obtained from an infrahuman animal (see p. 67).

Yerkes observed further and, perhaps more importantly, that "where very different methods of learning appear, the number of trials is not a safe criterion of intelligence." Thus *Julius* learned differently than monkeys, even more slowly, but his sudden change in behavior suggested "insight," rather than simple trial-and-error learning.

In box stacking experiments the orang exhibited slow, insightful behaviors but failed to solve the problem until its solution was demonstrated by the experimenter. When the boxes were removed, and the suspended banana could be reached only with a stick, *Julius* quickly solved the problem, doing so in unusually creative fashion.

> . . . instead of striking the banana with the stick and thus detaching it, he used the stick to climb upon or, thrusting it into the side of the cage, to swing out until he was able to reach the lure. Versatility and resourcefulness are terms which well describe the impression made on the observer by the animal's behavior. (1929, p. 190)

Yerkes and other early investigators were impressed with the manner in which orangs solved problems, despite the fact that their conclusions are admittedly more *subjectively* than *objectively* derived.

OBJECT MANIPULATION

In its natural habitat, the orang-utan uses the branches and vines of trees to facilitate its arboreal movement. As Wallace (1856) noted:

He walks deliberately along the branches, in the semi erect attitude which the great length of his arms and the shortness of his legs give him; choosing a place where the boughs of an adjacent tree intermingle, he seizes the smaller twigs, pulls them towards him, grasps them, together with those of the tree he is on, and thus, forming a kind of bridge, swings himself onward, and seizing hold of a thick branch with his long arms, is in an instant walking along to the opposite side of the tree. (p. 27)

Wallace also noted that when threatened, wild orang-utans climb to higher branches, breaking off quantities of the smaller ones, apparently for the purpose of frightening his pursuers:

It is true he does not throw them *at* a person, but casts them down vertically; for it is evident that a bough cannot be thrown to any distance from the top of a lofty tree. (p. 27)

A recent book by Campbell (1979) includes a very interesting description of orang-utan manipulativeness and I quote him here as follows:

Certainly the orang-utans . . . are among the smartest creatures on earth. Orangs, especially the young ones, are great favorites of zoo visitors. They have a sad clown quality in their expression, an affectionate nature, a sly playfulness, and a curiosity which, when it is combined with their intelligence and incredible manipulative ability, makes the problem of containing them at a zoo a constant challenge. They have fingers so dexterous that they can neatly field-strip a camera accidentally dropped into their enclosure by an unwary visitor, yet so strong they can twist off large nuts that have been tightened with a torque wrench. They seem to enjoy taking things apart, consequently any structure in an orang exhibit has to be strong enough to resist the leverage of their long arms and be joined by welds, rivets or bolts that have been countersunk (pp. 56–57).

In the wild, observations by Rijksen (1978) demonstrate the variability in manipulative mode of the Sumatran orang-utan. Not only the hands and feet are used since, as we have seen (cf. Chapter 1), the orang-utan possesses extremely mobile and sensitive lips.

In the manipulation of wild foods, orang-utans must face particular problems according to the nature of the food items. For example, many of the Durian fruits (Durio spp.) are encased in a spiny shell. Some of these plants can only be opened by humans with the aid of tools, and it is extremely difficult to carry them long distances without pain. According to Rijksen's observations, young inexperienced orangs do not carry or open Durian fruit so easily. The learning of proper processing is apparent in Rijksen's differential observations:

Younger orang utans usually collected only one fruit at a time. They rarely employed the method of pulling the fruit away from their mouth by means of their hands, as the adults did. Often they placed the fruit against a trunk, in a small crevice or in the fork of a branch, pushing it into place with their hands and then pulling out the spines one by one with their incisors. Once we observed that the sub-adult male Doba held a durian fruit into a small crevice by pushing it with a piece of dead wood, thus using a crude hand-protecting device. He had collected the tool some 15 meters from the place where he used it. (p. 84)

Interestingly, some rehabilitant orangs, having lived in captivity, regularly opened durians with the use of tools. Rijksen noted that *David* used paper, leaves, and on one occasion a gunny sack to protect his hands while holding spiney fruit. The orang *Usman* attempted to open fruit by using a pointed stick that he especially collected for that purpose. The stick action blunted the spines so that he could hold the fruit with greater ease, but it did not facilitate opening.

In the consumption of stinging colonial insects, Rijksen noted that orangs fed cautiously and attempted to occupy branches unconnected to the nest. Orangs generally regulated the numbers of emergent prey by enclosing a piece of nest material in their clenched hand or by enticing several insects onto the back of their hand:

Once we observed an orang utan tear off a twig containing an ant nest constructed of leaves presumably of an *Oecophyllus* sp. Grasping with his hand around the twig, just below the nest, the ape suddenly jerked towards the nest, thus stripping off a number of leaves from below the nest. These leaves apparently serve as a cover between the ants' nest and his hand. By holding this bundle of leaves, containing the nest in his clenched fist, he could regulate the number of escaping ants . . . picked off one by one with the lips. (p. 89)

Although termites were usually acquired in a conventional manner, rehabilitant orangs were observed to use the technique favored by wild chimpanzees at the Gombe Research Center in Tanzania (cf. van Lawick-Goodall, 1968). These orangs poked small sticks into termite tunnels, using the tool as a lever or probe.

An interesting example of defensive tool use was also suggested by Rijksen. In this case, a female with infant was attacked by a swarm of bees. She warded off the insects by waving twigs and pulling them over her body. As Rijksen pointed out, the shield of twigs gives credence to Brandes (1937) suggestion that the species-typical propensity to put objects on their head is derived from escape from stinging insects. Occasionally, rehabilitant orang-utans also poked sticks into rat burrows, but the purposes of this activity were not clear to Rijksen.

Aggressive object displays have been reported for both wild chimpanzees (van Lawick-Goodall, 1968) and wild and captive baboons (Kortlandt and Kooij, 1966; Maple, 1975; Hamilton, 1975). Rehabilitant orangs in the Rijksen study, when confronted with a caged clouded leopard, broke and dropped branches onto the cage. The female *Yoko*, which had been previously attacked by the leopard, actually tried to break into the enclosure, even poking long sticks at the cat through the wire.

In a series of recent studies (1976a, 1976b, 1976c, and 1977a, 1977b) Jurgen Lethmate has investigated the tool using proclivities of captive orang-utans. In one study, two young males were challenged by opaque boxes closed tightly with bolts. Lethmate studied tool-using and tool-making in three ways. First, he determined that tool use seemed to be "intrinsically motivated" since the two animals (2½ and 4½ years of age respectively) worked on the boxes with stick and rod tools without expectation of any visible reward. In a second test, the orangs had to perform up to 25 separate manipulations in order to obtain a visible food reward. The third test required that the subjects reach into a long box (which also had to be forced open) with still other tools. Since the animals generated tool-use in these second and third conditions including the "manufacture" of more efficient tools, Lethmate inferred the operation of a kind of "cognitive motivation" in these orang-utans. Interestingly, in a previously published report (1976a), one of the animals studied above (*Buschi*) learned to manufacture a double stick by chewing on the ends so that the tapered ends could be inserted into an iron tube. This tool manufacture spontaneously occurred in the absence of visible food reward, as above, but in nonreward contingent trials work proceeded more slowly and was punctuated with more behavior irrelevant to the task-at-hand. In this research, the ideas of both Köhler (1921) and Schiller (1957) have been supported. In the former, the idea is promulgated that relationships among components of a tool are insightfully discovered when a conventionally unattainable goal is confronted. Thus, Kohler's chimpanzees stacked boxes and then climbed them to obtain a suspended banana. Lethmate reports the same insightful behavior for his captive orangs, and Yerkes described the same behavior in the gorilla *Congo* (1927). Insight notwithstanding, Schiller was the first to demonstrate that chimpanzees do not need a seemingly unattainable reward to stimulate insight. They are apparently "predisposed" to put sticks together as they manipulate them in the absence of reward. This "innate" component in a complex behavior pattern suggests that the tool-making capacity of apes is a *natural* propensity which they share in common with human beings. As we have seen, orang-utans too are naturally inclined to manipulate and modify objects, and subsequently employ them in the further exploration and exploitation of their environment.*

Lethmate also reported the manufacture of drinking tools by orang-utans. Although MacKinnon (1974) reported that orangs rarely drink water in the wild, Rijksen (1978) asserted that the free-ranging and rehabilitant Sumatran orangs in his study area frequently utilized naturally occuring rain water "bowls" which were available in the rain forest. Captive orangs drink water regularly. To stimulate tool use, Lethmate placed a bowl of water beyond the cage bars in view of three young orang-utans. To acquire water, the subjects used sticks, wooden splinters, and leaves to reach into the liquid and then lick it off. An adult female and a five-year-old male used a stick to "fish out" a second leaf tool which had been placed in the water bowl. An especially "insightful" solution to the problem was discovered by the young male which stripped leaves from twigs, chewed them, and then used the leafy mass as an absorbent "sponge." The developmental aspects of such behavior are illustrated by the author's discussion of the behavior of his youngest subject:

> The youngest animal (3 years old) disposed of all single motor patterns which were necessary for such a tool manufacture. However, the animal did not integrate these motor patterns into a complete performance. The act of dipping a drinking tool into the bowl was not always successful, because evidently a precision grip had not been developed. It is supposed that the development of a complete tool-making performance is improved by trial-and-error learning. Observational and/or imitative learning could accelerate the development in wild and socially caged apes. (p. 251)

---

*In a personal communication, Lethmate has told me about an orang which used a tool (wood-wool) to "splint" a broken stick.

Finally, Lethmate addressed the behavior of nest-building which is a very functional behavior common to both wild and captive orangs (cf. Chapter 1). Lethmate's subject was an isolation-reared male orang-utan. In spite of its deprivation experience, this animal exhibited species-typical nest building, although with inappropriate objects such as blankets and food. This behavior was observed as early as the fourteenth month. Novel objects introduced into the cage (twigs, wooden blocks) were treated as nesting material, and even during formal problem-solving tests this animal attempted to employ various tools in the construction of the nest. Clearly, as Lethmate suggested, orang-utan nesting behavior is a good example of what ethologists have called a fixed-action pattern. Nonetheless, the propensity to build a nest requires further experience for it to be refined into the functional and complete nest-construction which characterizes the mature orang-utan living in nature.

This manipulative tendency has also been discussed by Gewalt (1975) who described the construction of a "rope" by captive orang-utans. Similarly, Jantschke (1972) notes the propensity of orang-utans to tie knots in suspended fibers and chains. Jantschke's research also determined a tendency for orang-utans to use objects during wiping/painting movements. Orangs investigated objects with the employment of such hand movements. In these instances they wiped not only walls and cage floors, but other objects and their own bodies. As in Lethmate's subjects, Jantschke's animals also used objects to absorb water and wipe it onto walls and other surfaces. Like chimpanzees, orangs in the Jantschke studies used objects to "paint" urine, saliva, and dung onto the various cage surfaces.

Manipulativeness has also been investigated in more systematic fashion such as the work of Parker (1969). In the author's own words:

> In the problem solving behavior of primates one finds a strange mixture of the brilliant and the backward—incredibly intelligent behavior interspersed with what appears to be a conceptual grasp of entire situations. (p. 160)

In Parker's paper, responsiveness, manipulativeness, and implementation behavior were the three basic response patterns investigated. In this first category, Parker provided access to a manipulable object under three conditions: visual access only, tactile

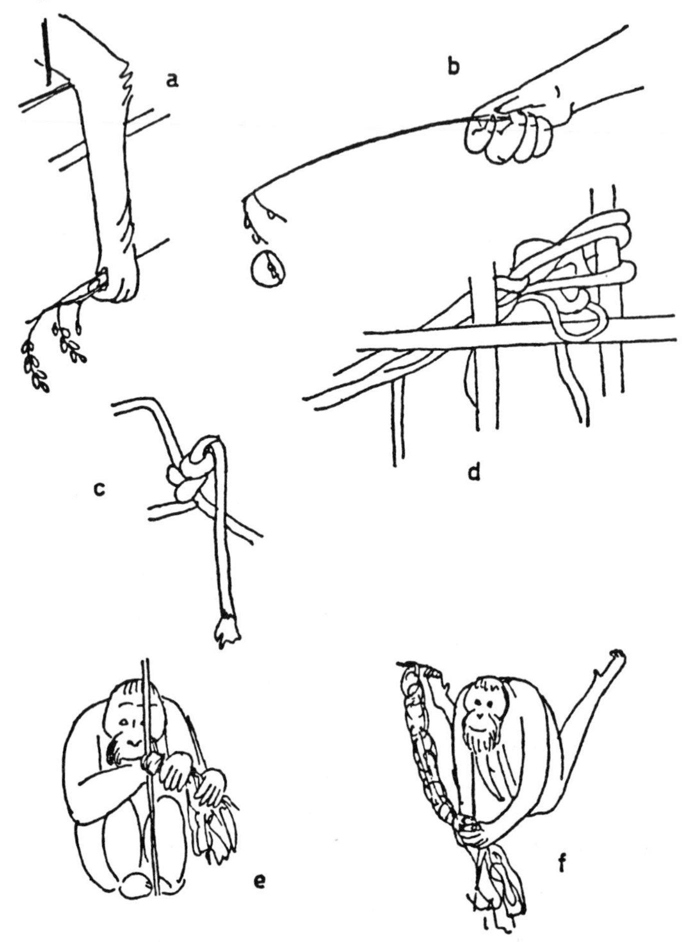

Figure 6-1. Knots tied by captive orangs (after Jantschke, 1972).

contant only, and both visual and tactile contact. Responsiveness was measured by the number and duration of contacts, and the *average* duration of contact per response. Comparatively (Figures 6-3a and 6-3b), orang-utans were more responsive than other species of great ape, although unlike gorillas, their responsiveness waned with time. Orang-utans were also more likely than gorillas to explore ob-

Figure 6-2. More knots. (After Jantschke, 1972.)

jects in the tactile rather than the visual mode. In the *quality* of manipulation Parker examined the following:

I. Body part utilized—fingers, knuckles, mouth, etc.

II. Object contacted—various details of the test object and apparatus.

III. Behavior displayed—A. Actions performed *on* the object—grasp, pull, bite, etc. (28 categories) B. Actions performed *with* the object—strike, scratch (nine categories).

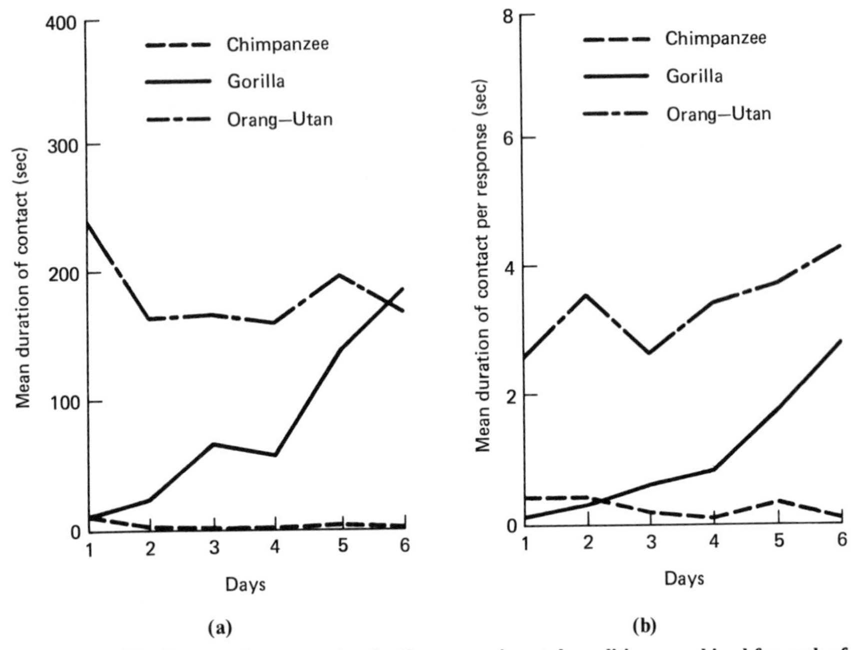

Figure 6-3a. Total responsiveness under the three experimental conditions combined for each of the three species as measured by the mean cumulative duration of contact per trial (maximum = 900 sec).

Figure 6-3b. Total responsiveness under the three experimental conditions combined for each of the three species as measured by the mean duration of contact per response. (After Parker, 1969.)

It was clear from Parker's analysis that orang-utans emitted a greater variety and number of responses than did gorillas or chimpanzees. While the initial response of orangs was typically a grasp, both chimps and gorillas cautiously touched with fingertips and backs of hands before grasping objects. Orang-utans were just as typically inventive in their use of the objects themselves:

The orang-utans were the only ones to twist the object a number of degrees greater than could be accomplished without releasing the grasp to make a second twist. They were unique in the use of the *pencil grasp* and in their attempts to push the object through the hold from which the tethering chain protruded. They singled out more details of the objects and apparatus to explore than did the other species. (p. 162)

With respect to *implementation*, as Parker used the term, two factors were identified as especially important: (1) the initial response repertoire of the subject relative to the task (motor aspect) and, (2) the abstractness of the relationships inherent in the problem (perceptual aspect). From here, Parker developed specific tasks with which to examine those factors as outlined above. Two such tasks, the "*Dip* problem" requiring the insertion of a piece of rope into a hole in order to extract banana mash, and the *Hoe* problem, in which fruit is drawn within reach by means of a J-shaped tool. While all species of great ape are known to have solved such problems, Parker's contribution was to relate problem success and mode of solution to the degree of responsiveness and the initial response repertoire of the subjects.

Parker's results on these two problems are summarized in Tables 6-1 and 6-2. On the *Dip* problem, the results again indicated that orang-utans performed more effectively than gorillas or chimpanzees. On the *Hoe* problem, orangs were again superior to their anthropoid cousins. Parker summarized performance into four categories as follows:

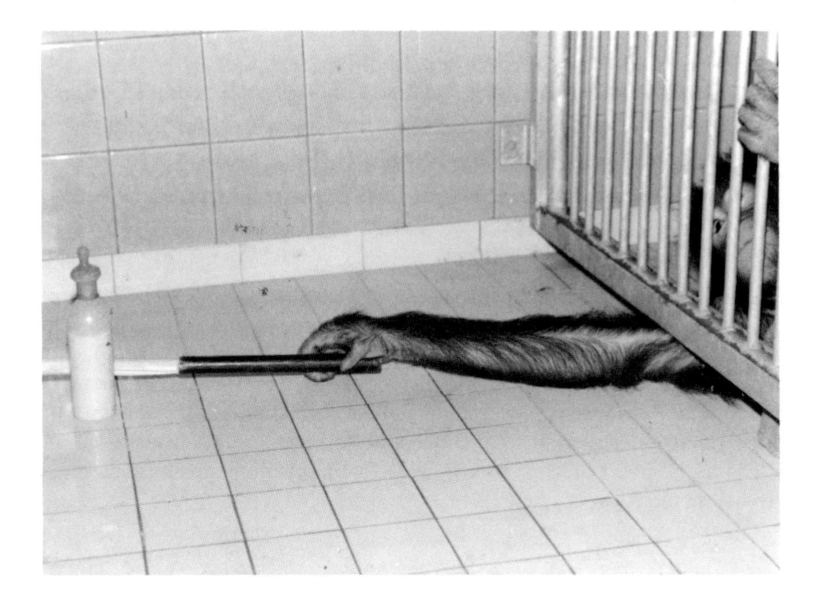

Figure 6-4. Orang-utan using stick as tool to obtain bottle (Photo courtesy of J. Lethmate).

1. *Immediate solution without trial and error or any display of irrelevant response patterns.* To all appearances they already knew how to solve the problem. The three subjects displaying this solution were orang-utans. They also displayed the appropriate behavior before the food incentive was introduced.
2. *Throwing and striking the hoe and striking the apparatus with the hoe.* This was clearly a running through of a portion of the response repertoire in the vicinity of the lure with solution being a matter of trial and error. The impression is that the behavior was directed at the incentive and the situation elicited only a part of the total response repertoire observed during the dry trials. The three subjects achieving solutions by this means were all gorillas.
3. *Solution only after final structuring, i.e., with the food in the crook of the hoe.* Attention was directed at the incentive and/or at the hoe but with no attempt to move the hoe toward the incentive. Two chimpanzees are represented here.
4. *No manipulation of the hoe and no solution*: one chimpanzee. Obviously, the quantity and quality of responsiveness and manipulation are highly related to behavior in this special situation, but the fine details of the correlations cannot be described here. In addition, there are additions in performance on two examples of the same type of implementation problem. It is thought that a significant factor within a problem type is the relationship between the motor patterns required and the motor repertoire of the subject, but that the major factor between problems is the perceptual relationships involved. A great deal remains to be discovered about the relationships between the motor and perceptual aspects of implementation behavior. (p. 166)

Manipulativeness may well be an indicator of higher intelligence. Although monkeys and other mammals are also manipulative, the propensity to do so is best developed in the apes and in human beings. Yerkes (1927) even suggested that manipulativeness was akin to science.

Fooling, in the sense of working with objects in a seemingly purposeless but exploratory fashion, is characteristic of certain types of mammal, and especially of the primates; characteristic also of certain periods of development, as for example human infancy and the early stages of many of the infrahuman primates. Presumably the more readily and persistently an animal fools with the objects of its environment, the more rapidly it learns their qualities and comes to adjust itself profitably to them. From one point of view what I term fooling or monkeying is the initial or primitive form of research! (p. 152)

Less conspicuous than in the monkeys, according to Yerkes, orang-utan fooling was said to be "more definitely controlled and pre-

Table 6-1. Performance on the Dip problem

Species	Name	Dry Day Sol.	Wet Standard position Day 2	Day 3	Day 4	Day 5—Structuring Trial 1 Standard	Trial 2 Remove	Reinsert	Trial 3 Reinsert
Chimpanzee	Aphie	No	2 min.						
	Dolly	No	—	—	—	—	Yes	No	No
	Gina	No							
Gorilla	Jr.	No	—	—	—	—	Yes	Yes	No
	Dolly	No	—	—	—	—	No	No	
	Timbo	No	—	—	—	—	No	No	
Orang-utan	Bob	Yes	3						
	Robta	Yes	2						
	Maggie	Yes	30						

Dry day: Yes or No refers to whether or not the rope was inserted into the hole.
Day 5: Remove = did the subject pull the rope out of the hole?
Reinsert = did the subject put the rope back into the hole if removed?
Trial 3 Reinsert = did the subject solve the problem after once removing the rope from the hole with food on it? Numbers in the cells indicate the time in minutes for solution. A minus sign indicates no solution, an empty cell indicates that the subject was not tested on that condition.

(After Parker, 1969.)

Table 6-2. Performance on the Hoe problem

Species	Name	Dry Day	Wet Standard position Day 2	Day 3	Day 4	Structuring Day 5 Tr 1	Tr 2	Tr 3	Day 6 Tr 1	Tr 2	Tr 3	Test
Chimpanzee	Aphie	No	—	—	—	—	—	—	—	—	9 min	4 min
	Dolly	No	—	—	—	—	—	—	—	—	12	1
	Gina	No										
Gorilla	Jr.	Yes	—	—	—	—	—	4				1
	Dolly	No	—	—	—	—	—	—	—	—	3	1
	Timbo	No	—	—	4							5
Orang-utan	Bob	Yes	3 min									
	Robta	Yes	4									
	Maggie	Yes	9									

Dry day: Yes or No refers to whether or not the hoe was extended all the way to the rear panel and subsequently drawn to the front of the box.
Test = one trial with the hoe in standard position presented 5 min after a solution at any point in the structuring.
Numbers in the cells indicate minutes for solution. A minus sign indicates no solution, an empty cell indicates that the subject was not tested on that condition. (After Parker, 1969.)

cisely directed; more systematic and hence more highly observational, investigative, and experimental." Such a lofty description as this could even be applied to Yerkes himself!

Parker's (1974) study of manipulative ability demonstrated distinct species differences in what the experimenter called *secondary actions*, those in which the manipulandum was applied in relation to an object, e.g., winding a rope around the body, striking an object with a rope. Great ape subjects exhibited more secondary actions than representatives of other species. With respect to mean percentage of secondary actions, orang-utans ranked higher than any nonhuman primate tested (cf. Table 6-3). Since Essock and Rumbaugh (1978) have noted that secondary actions are necessary for tool use,

Figure 6-5. Young orang-utan stacking boxes to obtain suspended food item. (Photo Courtesy of J. Lethmate)

Table 6-3. Species in rank order for mean percent of behavior classified as secondary

Species	%
Orangutan	35.59
Chimpanzee	28.73
Gorilla	12.81
Capuchin	6.25
Macaque	4.85
Gbbon	3.54
Lemur	2.84
Langur	2.05
Guenon	0.00
Spider Monkey	0.00

(From Essock and Rumbaugh, 1978, after Parker, 1974.)

it is not surprising that the supreme tool users, the apes, exceed monkeys in this capacity. The superiority of the orang-utan confirms the zoo lore regarding its propensity to manipulate objects.

Ellis (1975) reported tool use in three captive orang-utans at the Oklahoma City Zoo. The 12-year-old male *Lestunku* regularly places a tire under the water spigot in his cage, thereby catching water in the tire. He then carries the tire away to drink, sometimes using it to soak monkey chow. He also uses the tire for the consumption of dry monkey chow and fruit together, and frequently employs pieces of rubber and leaves as a sponge. Finally, various sticks and stiff foods (such as carrots) are used as levers, whereby they are forced into crevices to remove plaster or just enlarge a hole.

Oklahoma's 17-year-old female, *Lotus*, has frequently used objects in her cage to secure things that are out of reach. In this activity, she selects the appropriate tool from those which are less suitable, and she also is said to modify tools to a limited degree.

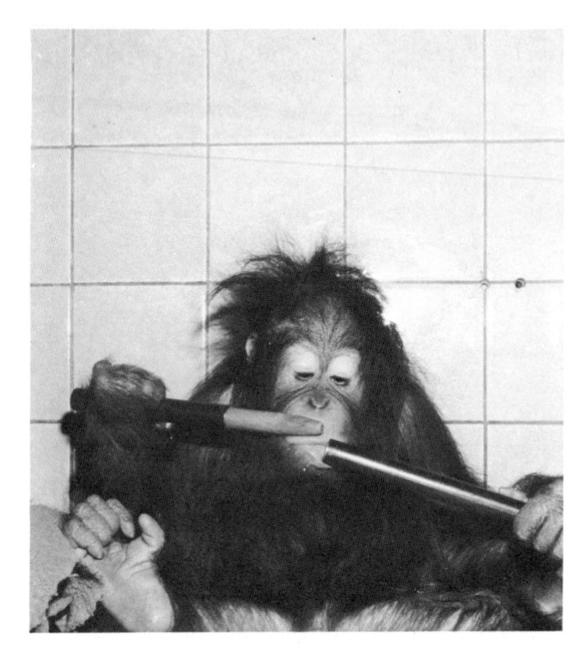

(a)

(b)

Figure 6-6a,b. Orang-utan modifying stick in order to fit pieces together to use as a tool. (Photos courtesy of J. Lethmate)

The 2-year-old male at Oklahoma exhibits tool use in threat when he brandishes a stick or branch toward the object of his dissatisfaction. He has also poked sticks in holes and sniffed and tasted the probe upon its removal.

Because of their inherent manipulativeness and ability to quickly learn, indeed, their apparent *need* to do so, many captive settings have employed devices whereby apes can operate their own feeders, solve problems for a reward, engage in physical activity, and interact with the public (cf. Chapter 7 for further discussion of these procedures). For example, Murphy (1976) described a feeder constructed at the San Francisco Zoo whereby chimpanzees can operate a connecting hose and levers to deliver food to themselves. As Murphy suggested:

> In its final form the feeder may allow for changes in the sequence in which the levers must be pushed, pulled or turned. Changes in the sequence could help to provide variety and challenge the problem-solving ability of the chimps.

Orang-utans were also given feeders at San Francisco. Here the device was placed above ground in order to reward the animals for climbing activity. Interestingly, the orang-utans often operated the system "simply to *watch* the food drop out." Accordingly, Murphy noted that the reward schedule of the device had eventually to be altered in order to preserve the food supply.

Perhaps the most important finding in the employment of this device was that the orangs climbed the structure 30% more often when the device was available. In addition, the zoo staff reported an increase in general activity and a reduction in stereotyped pacing by one female.

Similar to the San Francisco device is one installed at the Phoenix (Arizona) Zoo and studied by Schmidt and Killeen (personal communication). An interesting form of "altruistic" behavior developed during the course of the machine's use, according to Schmidt and Killeen's yearly progress report (1977, unpublished):

> Only Duchess (the seventeen year old Bornean female) was apparently willing to deliver pellets without obtaining them. Duchess also quickly became selective about to whom she would deliver pellets. She refrained from pressing the button when Joe (ten year old Bornean) was near the feeder drop area. Ben (18 year old Bornean) would occasionally sit below the feeder and Duchess would deliver several pellets in

a row into his lap . . . Duchess also delivered pellets to her two youngsters, though the intrusion of Joe often caused Duchess to leave the button.

We can confidently conclude from all of this that orangs are both manipulative and patient. In a relevent anecdote, the comparative psychologist Harry F. Harlow was quoted as follows regarding the perseverative nature of *Jiggs* the orang:

. . . the nicest and sweetest orang-utan that had ever lived at any zoo for fifteen years. We gave him two oak blocks, one with a square hole and one with a round hole, and a square plunger and a round plunger. He learned to put the round plunger in the round hole and the round plunger in the square hole, but he never learned to put the square plunger in the round hole. He worked incessantly on this unsolvable problem for six weeks and then died of perforated ulcers, but at least he died demonstrating a level of intellectual curiosity greater than that of many University of Wisconsin students. (Hahn, p. 74)

MEMORY

After an absence of six months I have found that my apes have forgotten nothing that I have taught them, although during my absence their course of instruction ceased entirely and they refused to do for others what I had taught them. Both the orang-utan and the chimpanzee have been able to learn the letters of the alphabet in order up to M. This is merely a demonstration of memory . . .

Furness (1916), p. 286

How good is the memory of the orang-utan? In a study by Robbins and Bush (1973), two orang-utans were compared to two gorillas and two chimpanzees during a complex discrimination task. Without belaboring the experimental details, the experimentors wanted to see if the number of irrelevant intervening trials would affect the subjects' discrimination performance in later tests. Although previous studies had demonstrated that great apes exhibit very high levels of information retention over long periods of time (cf. Medin and Davis, 1973), variations of intervention time and task had not previously been attempted.

In examining the results of the Robbins and Bush study, it can be seen that there were no great differences in the subjects' performance. In all subjects, as the number of irrelevant intervening tests between the initial discrimination and subsequent discrimination increased, the proportion of correct responses on the second discrimination decreased. According to the authors, this performance was not unlike that found for other nonhuman primates and even human subjects. Although this study included some further manipulations which are beyond the scope of this chapter, the memory process of the great apes, including orangs, demonstrates considerable similarity to that of humans.

CREATIVE RESPONSIVENESS

In a study by Rumbaugh *et al.* (1972), the hypothesis tested was that the cultivation of coordinated and constructive manipulation of various test materials would enhance *creative* responsiveness to new materials. One orang-utan and one chimpanzee were assigned to each of three conditions as follows:

I. Training on *various* tasks with *various* materials.
II. *Various* tasks (same as I) with *constant* materials.
III. No training, but free access to *various* materials.

An example of the type of training task was the "transport task" in which the subjects carried an object from one location to another within a room. Subjects in Group I were given a variety of objects to carry, whereas Group II subjects were given only one object.

After nine months of training, all subjects were tested with new materials under the same initial task conditions. Subjects were observed to determine whether they would make novel responses either during the tasks or in formal "responsiveness" tests.

The authors' interpretations are puzzling in that the results of the first responsiveness test indicated that Group I subjects were overwhelmingly (or should I say underwhelmingly) unresponsive. In a second test, however, the experimenter was not present during the test in order to remove possible social inhibitions. As the experimenters explained it:

> . . . Group I subjects during the relatively unstructured responsiveness test kept waiting for specific training signals to be given by the experimenter.

In the second test, conducted in an outside playpen, there was little difference between any of the subjects with the exception of one orang female Group I subject.

> . . . She was particularly creative. For example, she attempted to put the lid on the canister, interacted richly with the canister (placing a hand on her head in it for a total of 27 times), used the canister as a stool and packed various materials in the canister 19 different times. The materials which she placed in the canister included even small rocks and lengths of grass and sections of paper (torn by her from the paper sack) in addition to the various blocks provided for her. It is our interpretation that . . . she incorporated into her play and responsiveness well organized elements of her prolonged training in another setting with other materials. (p. 402)

Based on the orang's responsiveness, the authors concluded that creative responsiveness and creative play are in part a function of two kinds of experience: (1) training in the use of objects, singly and in relation to others; (2) the use of different materials and objects in functionally equivalent ways.

Clearly, this orang seems to have derived benefit from the training described, but we cannot be certain that its performance isn't more a reflection of individual differences, a common problem in ape research. In fact, some of the "novel" behavior of this orang, such as holding an object on its head, can be more parsimoniously interpreted as species-typical propensities.

One implication of this study is that orangs (and chimpanzees) may benefit from specific kinds of experience. Although we don't require such evidence to convince us, it may be said that this experiment demonstrates that complex organisms require complex experience in order to reach their optimum behavioral potential. Hence, we must construct complex habitats and provide novel stimuli and challenging interactive settings.

At the very least, it cannot be denied that orang-utans can be creative. Anyone who has carefully observed them will readily concur with such a conclusion.

ATTENTION TO CUES

As in many of the studies cited previously, orang-utans have been utilized to study general principles of learning. It was Robert Yerkes' early belief, of course, that all great apes should be utilized in this manner. When orang-utans and gorillas were first made available to scientists at the Yerkes Primate Center (first in Orange Park and later in Atlanta), their usefulness in studies of learning became apparent.

One such study was conducted by Draper and Menzel (1964) in order to determine the *size* of a reinforcer (food) sufficient to compensate for *distance.* The experimenters concluded that the selection of food varied inversely with distance, that is, nearer food was consistently preferred to more distant food. However, when the *size* of food was increased so that there was a difference of 0.10 in the food pieces, choice was determined by *size* rather than distance. The data finally indicated that as orangs reached an extra 5 inches for food, the diameter of the food had to be increased by 0.105 inches. Orang-utans behaved essentially as human beings do on such a task and they are clearly capable of perceiving small differences in size and distance (effort).

In related studies, Menzel and Draper (1965) gave orang-utans (and other primates) the opportunity to choose a standard size but concealed banana, or a variable size but visible banana. To no one's surprise, orangs exhibited a "bird-in-the-hand" mentality whereby they chose the *visible* reward under certain conditions. However, as in the prior experiment, other factors such as size can affect the choice. Orang-utans, as with other apes, are clearly capable of using a wide range of visual and mediated cues to food size. In this sense, as many experiments have indicated, apes form "expectancies" of rewards. Thus, given this cognitive ability, I have felt comfortable in asserting that the female *Sungei* (see Chapter 5) formed an expectancy about the ability of her newborn offspring (*Hati*) to climb.

An especially interesting study of cues was published in 1973 by Rumbaugh, Gill and Wright. Briefly, this study was concerned with the readiness of great apes to attend to visual foreground stimuli when they are relatively close to the eye. The relevant experiment demonstrated that orangs were more sensitive to the introduction of foreground cues (stimuli) than were chimpanzees or gorillas. This finding is best interpreted in light of the relative arboreality of the three species. Rumbaugh *et al.* explained the finding as follows:

> It might be that readiness to attend to foreground cues is an adaptation to protect the eyes from obstructions that would otherwise threaten harm as locomotion in trees occurs either through semibrachiation or climbing, as with orang-utans and chimpanzees. This same line of thought suggests that as the gorilla has come to be adapted for an increasingly terrestrial way of life, the selective force for such a perceptual propensity has weakened. (p. 188)

Alternatively, and in line with this argument, readiness to attend to foreground cues might assist arboreal animals in the *location* of branches and vines suitable to locomotion. Thus, falls might thereby be prevented, an equally powerful selective force.

The idea that differential perceptual and cognitive abilities can be best explained by phylogenetic adaptations is a theme worth stressing. Indeed, the behavioral propensities of the orang-utan, in every respect, can be understood only in reference to their arboreal way of life and the various physical adaptations which determine their lifestyle.

Rumbaugh and Gill (1973) have commented on this phenomenon further as follows:

> . . . it would seem that just as some people fail to see the wood for the trees, the orang-utan, more so than either the chimpanzee or the gorilla, might be unable to see the trees for the leaves. (p. 188)

PROBLEM SOLVING

In 1967, Rumbaugh and McCormack determined that the skills of the three great ape taxa were essentially equivalent. This conclusion was based on studies of learning set (LS) whereby animals learn a pattern or set by experience with a series of discrimination problems, indicative of a kind of "insightful" behavior.

In this same publication, Rumbaugh and McCormack looked also at oddity concept skills in the three great ape species available to them. In these tests, the subjects were required to determine the odd member of three objects. Performing at the 65–90% correct levels in these tests, there was little difference in the performance of chimpanzee, gorilla and orang-utan.

In a very interesting comparative study, Fischer and Kitchener (1965) found that orang-utans performed as well as either gorillas or chimpanzees on patterned-string problems. Thus, in these studies there is to be found little difference in the behavior of *Gorilla*, *Pan*, and *Pongo*.

In a very recent publication, Davis and Markowitz (1978) investigated the performance of a pair of orangs when given free access to a simultaneous light-dark discrimination problem. In a problem such as this one, the subject is given a food reward for choosing, say, the *light* stimulus. After a number of trials (in this case 20), a reversal occurs whereby only the *dark* stimulus is rewarded. The investigators reported a dramatic increase in responding beginning with the first reversal. In addition, there was a "marked shift toward decreasing numbers of responses required for reversals."

Davis and Markowitz' approach differed considerably from traditional ape learning studies. Instead of isolating their subjects, they carried out their research in a zoo enclosure, where the two animals could freely move about. This is an important difference in that it demonstrates that cognitive research can be carried out in zoo settings. Moreover, the experimenters did *not* deprive the subjects of food, a standard lab procedure, asking the animals to work for food even though food was freely available in the enclosure. Surprisingly, orang-utans will work under these conditions.

A primary motive for this research was the desire to provide stimulation for the animals, a focus of Markowitz' well known program of behavioral engineering. The data reported were acquired from the male only since he made over 99% of the responses. Thus, in an enclosure, dominance determines access to the apparatus. The results of the study do not greatly differ from other research (cf. Rumbaugh and Gill, 1971), except that the subject did not reach the point of responding to a reversal in one trial. As the experimenters argued: ". . . we believe that the complex milieu (the flow of zoo visitors, zoo routine, maintenance procedures, etc.) was an important determinant of the average criterion level."

Traditional learning research with anthropoid apes has greatly decreased in recent years. Davis and Markowitz' success at Portland's Washington Park Zoo should encourage further research in those settings in which the vast majority of the world's captive great apes reside.

RESPONSE TO IMAGES

There is very little information on the response of orang-utans to images on paper, mirrors, or film. However, Harrisson noted in *Ossy* at the Dresden Zoo that pictures of human faces elicited kisses on the eyes, nose, and lip areas of the photo. Moreover, *Ossy* exhibited a special interest in pictures of leaves and flowers which he poked at

Figure 6-7. Anthropologist Lyn Miles with signing orang-utan *Chantek*. (T. Maple photo)

and tried to chew. He did the same whenever he encountered clothing which was marked with flowered print. We have shown our orangs pictures of other orangs and they have exhibited considerable interest in them.

Jürgen Lethmate and Gerti Dücker (1973) have demonstrated that orang-utans, like chimpanzees, apparently recognize themselves in a mirror. When they are marked with dye, they respond to their image by touching their own head. Schmidt (1878) reported an orang's first response to a mirror in which the animal approached slowly, exhibited piloerection, protruded his lip and then retreated. After another approach and swift withdrawal, the orang spat upon it, and threw in its direction a wooden hammer and crusts of bread. Later, the orang appeared to invite the image to play by rolling a ball in its direction. Contrary to Lethmate and Dücker's later findings, the author found that this orang-utan did not recognize himself. As Gallup

(1975) has shown with chimpanzees, self-recognition varies with social experience. It may be safely concluded that orangs *are* capable of the perceptual distinctions which seem to separate both man and ape from monkey.

As for motion picture images, little is known. We have only the anecdote that the Sacramento orangs (cf. Chapter 7) responded to their viewing of gorilla sex films by producing an infant. Perhaps they are even more perceptive than we imagined. The more "real" the image, the more likely it is to elicit responding. I would suggest here that holographic stimuli would be likely to stimulate considerable emotional response from any apes to which they were presented. The use of such stimuli in experimental research doubtless would be most successful.

ORANGS AS ENTERTAINERS

Bobby and Joan Berosini advertise their act as the "only trained orang-utan act in the world." It is the only act of its kind known to this writer. For the Berosinis it is an act with a long history, since Bobby's father trained orangs in Europe. The act is now based in Las Vegas, employing both a male and female orang and several chimpanzees.

The Berosinis have explained the key factors in their successful training as (1) an emphasis upon "communication" with their animals, (2) long, daily work sessions, and (3) the training of only those behaviors which occur naturally in the orang-utan repertoire.

The last factor is especially important. Learning theorists have repeatedly demonstrated since the work of Keller and Marian Breland (1962, 1966) that built-in constraints permit the easy acquisition of some responses, while others are difficult or impossible. An example of this is the ease with which Köhler's (1925) chimpanzees were able to put together interchangeable segments of a three piece pole. Köhler suggested that this behavior was an example of insight. However, as we have seen, Schiller (1951) argued that chimpanzees "naturally" put together sticks in the absence of a need to solve a problem. In Schiller's view manipulativeness was an innate propensity of the chimpanzee. Similarly, tricks or problems which require the insertion of keys into locks should be easily learned by an animal which fishes into mounds for termites with small sticks (cf. Goodall, 1968).

In elaborating upon the communicative aspects of orang training, the Berosinis do not use food rewards with these animals, but rely on tactual reinforcement and social approval. With experience, the animals cue on subtle bodily and verbal commands, and respond to please Bobby who is the "alpha" member of the group.

Two of the more humorous behaviors in the act are "kissing" and "smiling." Both of these responses occur commonly in the repertoire of the orang. The former is a type of greeting, and also occurs during food-sharing. The latter has been described as the fear grimace (cf. Rijksen's *horizontal bared-teeth* face) and occurs frequently in captive animals. The Berosinis have essentially exploited a response which is naturally* emitted to a superior, shaping it to occur on command. People who observe the grimaces of monkeys and apes universally interpret it as a grin.

Perhaps the most remarkable thing about the Berosinis' training methods is that they have successfully worked with an animal which others have found intractable. Certainly, the stoic temperament of the orang requires more patience on the part of the trainer. Their responsiveness to training make it clear that the limits of the animal's ability to learn have not yet been reached. As remarkable as these accomplishments may be, a contrary attitude to public entertainment has been offered by Harrisson (1962):

> Orangs, in contrast to chimps, are not good entertainers . . . their minds are not set to imitate, but to explore. Why is it that zoos and circuses insist on the idea of "teaching" them to sit on a chair or round a table? Would not a much greater crowd be drawn if they were let free in the trees of the zoo to show the spectators how they can move, climb, build nests? (p. 136)

LANGUAGE ACQUISITION

When I first became aware of the work of the Berosinis, I was convinced that the orang-utan's reputation as a student had been under-

---

*The Yerkes (1929) also commented upon the requirement of responding to the orangs natural propensities as follows:

> . . . play and exercise should be devised in a way as to appeal to the animals' way of thinking and inclinations; not to that of his trainer. (p. 136)

estimated. Indeed, scientists are now engaged in attempts to teach the orang-utan sign language. *Pongo* is the only great ape genus which is not currently represented within the linguistic sorority of *Koko* gorilla, *Washoe*, *Lana*, and *Sarah* chimpanzee, *et al.* In California and Tennessee, orang-utan signing is now being shaped. I will discuss here the situation with which I am most familiar, Tennessee's famous *Chantek*.

Initiated in 1978 by Dr. Lyn Miles of the University of Tennessee's Chattanooga campus, the male *Chantek* was obtained on loan from the Yerkes Regional Primate Research Center. Interestingly, the California project employs a young male orang (*Bulan*) also on loan from Yerkes. *Chantek* was acquired at one year of age and he, too, is in constant contact with his human caretakers. When they are young, as we have seen, orangs are very active and not altogether unlike young chimpanzees in temperament. If they are to learn, young apes require the same perseveration of the teacher as do human infants. The biggest obstacle to training is their short attention span. Like the Berosinis, Professor Miles has worked hard to establish a communicative rapport with her subject. As this book goes to press, *Chantek* understands several signs but has not yet learned to sign himself. Because he has responded well to the training situation and in view of the Berosinis' success, it is likely that *Chantek* will be just as linguistically prolific as any of the other apes. The orang signing data should move us closer to a comparative analysis of language skill.

One of the most fascinating aspects of *Chantek's* linguistic potential is that he has exhibited a propensity to sign with his feet. This should be considered one more reflection of the arboreal adaptations of the species.

Furthermore, in consideration of the orang's sexual proclivities, it would be interesting to teach the infant a sign for genital-contact (cf. Chapter 4), in order to assess contact motivation relative to food, etc.

An interesting historical note is that Furness (1916) once attempted to teach an orang-utan to speak. In so doing, Furness reported that orangs were surprisingly capable of *understanding* human speech, but it was not an easy matter to train them to talk. Succeeding in teaching the animal to utter the words "cup," "papa," and the

Figure 6-8. Animal trainer Bobby Berosini with trained orang-utan. (Photo courtesy Bobby Berosini)

sound "th," the experiment ended when the student died. The techniques used by Furness were described as follows:

> The training consisted of a repetition of the sounds for minutes at a time, while the ape's lips were brought together and opened in imitation of the movements of my lips. I also went through these same maneuvers facing a mirror with her face close to mine that she might see what her lips were to do as well as feel the movement of them. At the end of about six months . . . she said "Papa" quite distinctly and repeated it on command. Of course, I praised and petted her enthusiastically.

Clearly, studies assessing the cognitive capacity of orang-utans must utilize a form of language which is readily learned. So far, computer systems, physical symbol systems, and sign language seem to be modes of communication ideally suited to the sensory and motor abilities of the orang. However, if the legends of the Dyak people are correct, orang-utans can already talk, but they do not want us to know about it for fear that they will be made to work. Now, that would be an indication of intelligence!

AN INTELLECTUAL SUMMARY

So how smart are these red apes? On a variety of tasks, the orang has proved to be equally as bright as the chimpanzee and gorilla. In its natural tendency to perserveration and manipulation it may even exceed these two in its ability, as recent evidence indicates.* However, if we look to the Yerkes for an early comparison, we see that, as with *temperament*, they have suggested 12 different categories of intellect. I have organized and summarized these comparisons in Table 6-4. In inspecting this table, the reader will note that there are two rating summaries, one which the Yerkes *suggested*, and one which may be regarded as the *true* ranking. In the Yerkes' opinion, a combined order was not clearly indicated by the information available to them in 1929. Thus, using subjective personal experience to assist them, they

---

*As Essock and Rumbaugh (1978) have argued, equivalence of performance among the great apes is contrary to zoo lore.

> Chimpanzees are often reputed to be the "smartest" of the apes, and orangutans have the reputation of being dull and sluggish. Such tags are unfortunate and contrary to the results of studies presented here. Much of the lore probably stems from the chimpanzee's tendency to mimic (ape) human behaviors. We consider ourselves marvelously clever; ergo, so is the chimp. Gorillas poke things with their fingers that chimps slap with their palms. And orangutans will, likely as not, pick at the screw *next* to the reward. At the end of a test session, an orangutan or gorilla may be slowly dissolving a mouthful of M & M candies, while a chimpanzee will not. Such variations in temperament and distractibility can lead to erroneous conclusions about the relative *capabilities* of these animals (p. 185).

concluded that the gorilla, rather than the chimpanzee, was the closer to humankind in cognitive skills. As they were inclined to qualify:

> This we grant is not justified by the ratings of the preceding paragraphs. Our explanation of what must seem a discrepancy is that the chimpanzee has been much more extensively and carefully studied than any other anthropoid. Were correspondingly varied and satisfactory data available for gorilla, it might command first place in psychological resemblance to man even if less alert, docile, imitative, and suggestible than the chimpanzee. (p. 550)

The same, of course, could be said of the orang-utan today. We are just now beginning to understand the red ape in both captive and field conditions. Perhaps its rank could be elevated if more data were available. However, it should be understood that each of the apes is uniquely suited to its habitat, and its emergent life style and behavioral adaptations reflect the pressure of thousands of years of natural selection. Thus, as Hodos and Campbell (1969) have asserted, a linear order of intellect does not accurately reflect either the relationship of the respective animals, one to the other, or the relationship of each to its characteristic habitat. Orang-utans are slow and cautious in the trees, and they are slow and cautious in the manipulation of objects in laboratory and zoo. Orangs are structurally suited for

Table 6-4. Composition comparison of intellect in *Gorilla, Pan,* and *Pongo.* (Adapted from Yerkes and Yerkes, 1929)

	Orang-utan	Chimpanzee	Gorilla
Curiosity	2	1	2
Imitation	2	1	3
Tuition	2	1	2
Attention	3	2	1
Adaptation	3	1	2
Memory	2	2	1
Imagination	2	1	1
Instrumentation	2	1	3
Tool use	2	1	3
Manipulation	2	1	3
Adaptivity	3	1	2
Resemblances to Man (Yerkes)	3	2	1
True Rank	3	1	2

easily grasping branches, and both their play and their manipulative propensities reflect this adaptation. Thus, we can better understand their temperament and their intellect in view of natural selection and the characteristics of their ecological niche.

Harrison has summed up the general abilities of orang-utans as follows:

> The orang's natural curiosity and keen sense to explore, his slow contemplating mind with a capacity to remember events, make him *a priori* highly successful in the jungle. But these very attributes also make him successful, especially as a juvenile, in *human* surroundings . . . " (1962, p. 82)

7
The Captive Habitat

"Orang-utans are skeptical of changes in their cages . . ."

At the Zoo, Simon and Garfunkel

In this chapter, I will review the state of the orang-utan in captivity, and offer recommendations for their proper management. Many of these statements apply equally to gorillas and chimpanzees and are based on a previous chapter published in Erwin, Maple, and Mitchell (1979). It is my hope that the ideas presented here will be of value to those who carry the special responsibility of ape-keeping. In a more general sense, these points and principles will very often apply to the management of other species*, and I trust that they will be of similar utility. In preparing this contribution, I am in agreement with Van Hooff (1973) in acknowledging "a growing feeling in the last few years that the conventional methods of keeping great apes . . . are not fully adequate." The Yerkes' pointed out, however, an area of disagreement in discounting the view of Köhler who, like Van Hooff, worried about the environments of captive apes.

> Köhler (1921, p. 73) in emphasizing the essentially social nature of the life of the chimpanzee, has quite appropriately questioned the naturalness of the adaptive behavior of this ape in isolation, and extending his generalization he has suggested that experimental studies of the behavior of the isolated orang-utan may yield misleading results. The content of orang-utan literature would suggest on the contrary its relative independence of the factors of social environment and its ability to live contentedly and adapt itself with the usual measure of success in nature or in captivity as an isolated individual. (Yerkes and Yerkes, 1929, p. 140)

What then are the relevant variables which affect orang-utan adaptation? And how does the red ape fare in captivity?

---

*Habitat design has become an increasingly important topic in literature of psychology, leading to improvements in the quality of human environments (cf. Sommer, 1974 and Ellenberger, 1960, for direct comparisons of animal and human habitat requirements.)

A BRIEF HISTORY

While both chimpanzees and orang-utans were exhibited in Europe as early as the eighteenth century, few specimens lived long. Indeed, the failure to observe apes for any appreciable period of time, in the wild or in captivity, led to considerable confusion over the number of extant species. A common misconception was the proposition that young female apes were "pygmy" or "dwarf" species. Furthermore, the confused taxonomy of the 18th, 19th, and early 20th century was an obstacle to an accurate historical record, since apes were often inaccurately portrayed as representative of other species. For example, Lang (1959) has argued that the first recorded exhibition of a gorilla (1855) was, in fact, that of a chimpanzee.

In the public display of orang-utans, as with all primates, the major obstacles to survival were their susceptibility to human disease, inadequate diets, irregularities of climate, and, undoubtedly, the effects of stress. In this century, the earliest breeding and housing successes for apes were recorded by Madame Rosalia Abreu in Cuba. Madame Abreu's colony at Quinta Palatino was described by Robert Yerkes in his 1925 volume *Almost Human*:

> If, then, we were asked to sum up for the mistress of Quinta Palatino, as well as ourselves, the essentials of success in keeping and breeding the higher primates, we should emphasize the following points: freedom, or reasonably spacious quarters, fresh air and sunshine, preferably coupled with marked variations in temperature; cleanliness of surroundings as well as of the body; clean and carefully prepared food in proper variety and quantity; a sufficient and regular supply of pure water; congenial species companionship and intelligent and sympathetic human companionship, which, transcending the routine care of the animal, provides for the development of interest if not friendliness; and, finally, adequate resources and opportunity both in company and in isolation for work and play.

In nearly every instance, as we will see, this advice is sage indeed. Yet, with few exceptions, the captive display of great apes has remained, at worst, a public disgrace, and, at best, a relic of less enlightened times. Elsewhere I have portrayed our *entire* history of apekeeping as an example of the good, the bad, and, *all too often*, the ugly (cf. Maple, 1979).*

In summarizing the accomplishments of Madame Abreu, Yerkes added that: "Given these conditions of captive existence, primates originally healthful and normal should without difficulty be kept in good condition of body and mind and should naturally reproduce and successfully rear their young." We have come to judge our success in ape-keeping in exactly this fashion: *successful bearing and rearing of offspring*. The psychologist, however, may well take a dim view of the narrowness of these two objectives. Our research into animal psychopathology (cf. Berkson, 1968; Erwin, Mitchell, and Maple, 1973; Maple, 1977; Rogers and Davenport, 1969) has broadened our understanding of stereotyped motor acts, autoerotic and autoagonistic behaviors. Therefore, we may wish to add that only animals free of these bizarre behavior patterns should be considered adequately housed and/or reared.

In reviewing the breeding and rearing success of orang-utans it is encouraging to note that some progress has been made. The first captive orang-utan birth was recorded in 1929, while by comparison gorillas did not reproduce in captivity until 1956 (cf. Bourne and Cohen, 1975). In all of the great apes, second generation births are the most difficult to achieve. (cf. Chapter 5). It is with these captive-born animals that rearing problems are also most apparent, for once the problem of breeding has been solved, the mother must then exhibit proper parental care.

Differential breeding success in orang-utans can be attributed to a number of factors, e.g., diet, habitat, social history, and group size. It is also clear that there are differences in the amount of sexual behavior normally exhibited by individual orang-utans (cf. Nadler, 1977), and in relation to this, captivity may differentially affect reproductive behavior in individuals as with species (cf. Maple, 1977; Maple, 1979). To properly plan for appropriate ape-keeping procedures, it is first necessary to understand their specific habitat needs in relation to species-typical physical and social factors.

---

*The sterility of the "ugly" captive habitat has been well described by Harrisson:

Here the apes' fun and games of the day consist of begging for food and performing tricks, such as spitting and urinating at spectators and clapping of hands (Chimpanzees); begging for food or staying hidden for hour after hour in a bunch of horse straw (Orang-utan); sitting stoically in front of the visitors for hours on end without moving (Gorilla). As hardly any visitor (and seldom a zoo director) has ever observed apes in the wild, there can be little conception of how far this exhibit is a distorted farcical picture of what once were apes, long ago and far away, in their homelands. (1962, p. 134)

As we have seen, the wild orang-utan has often been described as a solitary creature (cf. Rodman, 1973). However, such a description does not accurately portray the full social capacity of this ape. Clearly, the traveling habits or orangs are different from gorillas and chimpanzees. Mature orang males are generally found alone, and they maintain large territories which they apparently defend from neighboring males. However, as orangs have also been observed during prolonged consortships with females (cf. MacKinnon, 1974), even in the wild, they are capable of strong social attachments. In this writer's opinion, the social relations of wild orang-utans have yet to be fully described, despite the notable and careful efforts of Davenport (1967), Rodman (1973), Horr (1977), MacKinnon (1974), Rijksen (1978), and Galdikas (1979). As will become clear, the prevailing view of the orang-utan as a lethargic, solitary, emotionless creature has become a self-fulfilling prophecy in captivity.

What no one can deny about the orang-utan is that it is the most arboreal of the great apes, and whether in captivity or in nature, its movements are cautious indeed. As has been noted, the animal appears calculating in everything it does. Researchers who have attempted to assess its intelligence find the orang's temperament to interfere with its performance (cf. Chapter 6). However, with patience and hard work, it is possible to train orangs to emit an incredible array of behaviors. To emerge from lethargy, however, captive orangs require stimulating social and physical surroundings.

SPATIAL REQUIREMENTS

In considering the spatial requirements of orang-utans it must be acknowledged that no captive habitat can replicate the spatious dimensions of their natural habitat. Indeed, the construction of such vast enclosures would be impractical and prohibitively expensive. However, it is possible to provide environments which stimulate the same activities which would occur in the wild. For all of the apes, an important habitat dimension is the vertical component. Since orang-utans are overwhelmingly arboreal, they require elevated pathways in order to locomote in their characteristic fashion (brachiation). Habitats which do not permit brachiation contribute to the lethargy which often characterizes this species in captivity. Orang-utans are known to construct sleeping nests, and they will spend many hours forming these structures if they are given branches, hay, or some other form of browse. This activity is not only good for the animals, but is quite interesting to the viewing public. Platforms or raised sitting areas provide the opportunity for elevating these sleeping nests and this also contributes to activity.

By attending to vertical as well as horizontal space, the entire enclosure can thus be better utilized. Climbing and sitting apparatus increase the complexity and therefore the quality of the environment. In addition, where groups of animals are housed together (a preferred state of affairs), an increase in spatial volume can reduce crowding and subsequent social stress. Moreover, proper internal construction will take into account the need for *cover*. As Erwin et al. (1976) have pointed out, cover can effectively remove an animal from view, thereby providing a means to shorten conflict. Modified cement culverts, protruding walls, and room partitions are especially effective in providing refuge and privacy. Open sleeping dens have also been employed in this manner. However, one of the hazards of building new habitats, and even of adding new furniture, is that, with time, apes become attached to the characteristics of their enclosures. New surroundings or, more accurately, separation from old surroundings can induce stress, lead to illness, and even death. Older apes are especially susceptible to sudden changes in their enclosures, so improvements should be made gradually and with caution.

FLIGHT DISTANCE

Of particular importance in the construction of captive habitats is the requirement that the animals be given adequate spatial separation from human intrusion. While it is difficult to determine what an adequate distance may be, the better habitats which I have observed provide *at least* 20 feet of separation from the public as measured from the front of the cage. Of course, an enclosure with adequate depth will allow the animal to establish its own minimum distance. As Hediger (1950) has written:

Since flight reaction is the most significant behavior pattern of the wild animal's life in freedom, it must be a prime concern in captivity to give normal play to this vitally important reaction. This means giving the animal the chance to get away from man himself, at least to beyond its flight distance. The smallest cage in theory must thus be a circle of a diameter twice the flight distance.

WORK

In the wild, apes must travel great distances in order to acquire sufficient sustenance. Foraging for food is a kind of work in which all wild animals must engage. To encourage such exercise, it is necessary to devise ways of stimulating activity. Yerkes (1925) recognized the importance of this activity in his book *Almost Human*:

Undoubtedly, kindness to captive primates demands ample provision for amusement and entertainment as well as for exercise. If the captive cannot be given opportunity to work for its living, it should at least have abundant chance to exercise its reactive ingenuity and love of playing with things . . . The greatest possibility of improvement in our provisions for captive primates lies in the invention and installation of apparatus which can be used for play or work.

Captive apes need activity in order to prevent boredom, and promote health. Without such opportunities they will engage in unhealthy behaviors such as excessive self and social grooming, repetitive regurgitation of food, and coprophagy. The latter two behaviors are, in part, the result of the reduced food intake which is necessary in order to prevent obesity. Were the animals active, restricted food intake would be unnecessary and caprophagy and regurgitation would be unlikely to develop. Orang-utans do not consume their own feces in the wild, hence, as Hill (1966) asserted, "coprophagy is typically a behavior pattern of captivity." In his review of the literature, Hill cites a paper by Stemler-Morath (1937) in which coprophagy was linked to a curiosity-motive. Thus young apes which examine and then play with fecal material eventually learn to eat it. Boredom apparently facilitates the acquisition of the habit. Moreover, Stemler-Morath observed that when apes at the Basel Zoo lived in outdoor summer quarters they climbed trees, and ate leaves, bark, and large quantities of earth. Inside during winter, the animals resorted to urine and feces consumption. To alter the behavior, the author suggested the establishment of regular feeding times, immediate disposal of feces through water-flushing, opportunity for play, proper diet (especially in winter) consisting of browse,* and in some instances punishment. Hill also mentioned that coprophagy could, in some instances, be due to dietary deficiencies which are physiological rather than psychological. However, in many of the cases reviewed by Hill, an important factor seemed to be onset of stressful events such as loss of a cagemate or a change in regimen. One of the most promising solutions to such problems is the technology developed by Hal Markowitz (1979). These devices can induce locomotion, cooperation, and problem-solving efforts when designed properly. An interesting example of an especially creative (yet simple) device is a tug-of-war which operated at the *Honolulu Zoo.* (cf. Maple, 1979).

With this device a gorilla is given exercise and social stimulation by pulling a rope against the observer. Moreover, with this device the public is directly exposed to the strength of a gorilla. Games such as this and those described by Markowitz allow public-animal interactions which are safe, indeed, beneficial for the animal.* These innovative techniques clearly fulfill a need which has previously been satisfied through public feeding, petting zoos, and animal rides.

Hill recorded the view of Desmond Morris that:

> . . . coprophagy may be the result of the fact that, in the wild, apes spend many hours obtaining their food, whereas in captivity feeding behaviour is reduced to a few short, fast meals, leaving them with a great deal of 'empty feeding time'—and coprophagy is almost the only kind of feeding that is available to them. (Hill, 1966, p. 255)

With this explanation I enthusiastically agree. The solutions offered by Morris and others are to provide increased, essentially *ad libitum*, feeding opportunities, particularly if it requires foraging, e.g. small seeds. An additional benefit is the introduction of novel objects

---

*See Table 7-1.

*I fondly remember my first "contact" with an orang-utan. While observing the adult Sacramento Zoo pair, I noticed that Josephine was holding a wet muslin rag in her hand. She stoically peered at me, glancing at the rag and then back to me repeatedly but slowly. I knew that something was brewing in her head. Calmly she tossed the rag some thirty feet in my direction, whereupon I caught it in midair. I threw it back, and after a short inspection of the rag she threw it to me again, but this time it fell short and disappeared into the moat. It was a thrilling game which she and I might well have continued for some time.

Table 7-1. List of shrubs and trees cut for browse at the Philadelphia Zoo. (After Snyder, 1976)

1. Alder (*Alnus* sp.)
2. Beech (*Fagus* sp.)
3. Birch (*Betula* sp.) only in spring
4. Bush Honeysuckle (*Lonicera* sp.)
5. Butterfly Bush (*Buddleia* sp.)
6. Cottoneaster (*Connoneaster* sp.) in winter
7. Dogwood (*Cornus* sp.)
8. Elaeagnus (*Elaeagnus* sp.) in winter
9. Elm (*Ulmus* sp.)
10. Firethorn (*Pyracantha* sp.) in winter
11. Forsythia (*Forsythia* sp.)
12. Hackberry (*Celtis* sp.)
13. Hazelnut (*Corylus* sp.)
14. Ilex (*Ilex* sp.) in winter
15. Japanese Pagoda Tree (*Sophora japonica*)
16. Jasmine (*Jasminum* sp.)
17. Kentucky Coffee Tree (*Gymnocladus dioica*)
18. Kerria (*Kerria* sp.)
19. Linden, Basswood (*Tilia* sp.)
20. Mahonia (*Mahonia* sp.) in winter
21. Maple (*Acer* sp.)
22. Mock Orange (*Philadelphus* sp.)
23. Mulberry (*Morus* sp.)
24. Oak (*Quercus* sp.)
25. Poplar, Cottonwood, Aspen (*Populus* sp.)
26. Privet (*Ligustrum* sp.) in winter
27. Prunus (*Prunus* sp.)
28. Raspberry, Blackberry (*Rubus* sp.)
29. Robinia (*Robinia* sp.)
30. Rose (*Rosa* sp.)
31. Snowberry (*Symphoricarpas* sp.)
32. Wisteria (*Wisteria* sp.)

for play. Here again, barren, unchanging environments contribute to the problem. From Hills's review and from our experience, orangutans are somewhat less prone to feces eating, but commonly engage in urine consumption. As with all apes, orangs benefit from the same suggestions mentioned here and do not require any unique treatment.

One solution to coprophagy which has not been attempted is "learned taste aversion" as successfully utilized to alter the feeding habits of wild coyotes (cf. Gustafson, 1978). Hill (1966) describes a relevant case reported to him as follows:

When defaecation occurred, the keeper rushed into the cage and sprinkled bitter aloes on the faeces. This resulted in the chimpanzee defaecating into its hand instead of onto the floor.

The clever adaptation of the chimp serves to point out the problems inherent in ape management. However, the taste aversion technique requires the treatment of the undesirable food-object with a stronger substance such as lithium-chloride. This substance induces nonlethal sickness and vomiting which is generally effective in one application. The association of taste and sickness is quickly acquired and the effect is enduring. Under the supervision of a competent veterinarian, the procedure might well succeed in difficult cases of coprophagy. For details of the procedure and theory behind it the reader is referred to Bolles (1975), Green and Garcia (1971 and Gustafson et al. (1978).

It is interesting to note that the voluntary regurgitation of food is an exceedingly common behavior pattern among apes at the Yerkes Primate Center and the Atlanta Zoological Park. It is especially common among orang-utans. Equally distasteful to the viewing public, its origins are likely to be the same as in coprophagy and its solution as equally straightforward. In my opinion, there is no good reason for zoos to tolerate either. Its prevention requires only intelligent habitat design and effective management, however, its treatment may require an even greater effort including the destruction or modification of the current environment. As Hill concluded:

. . . ape exhibits should be constructed with coprophagy in mind . . . providing apes with sufficient activity while they are in captivity. Devices for climbing, swinging and other pastimes designed to occupy them should be offered to intrigue the apes' mentality, bearing in mind that even these can become boring and should be rotated to present new problems when necessary. (p. 256)

A sad story of the outcome of public feeding can be found in Benchley's book *My Friends, the Apes* (1948):

Jiggs became more and more childlike as he grew weaker; there was little left of him except the distended abdomen and the climbing fingers. He whimpered imploringly to each of us and, although we could do nothing for him, he loved to have us hold him in our arms. When one morning Henry came to tell me he was gone, I felt relief that his suffering was over. Autopsy revealed that he had a big wad of gum, paper and all, in his stomach. How he obtained it under Henry's watchful eye, we never knew. He could not rid himself of it as an older, stronger ape might have done.

Unlike public feeding, a game permits the zoo to exert control over the nature of the food-item. For the animals that are provided in-cage manipulanda, it is advisable to change the game from time to time. In fact, where any objects are introduced for play, it is helpful to substitute *novel* objects periodically. As has been previously emphasized, a *changing* environment is to be preferred over a static one. An added benefit of many games is that the cognitive abilities of the apes can be studied in the context of entertainment. Morris (1959) has pointed out the value of testing chimpanzee intelligence in view of the public, and Reynolds (1967) has further extended this view to include other forms of training:

> . . . young tractable apes can be used in training programs, circus-style, to brighten their lives and amuse the public. There is absolutely no reason why zoos should be shy of using circus techniques. These may result in healthier animals and greater subsequent breeding success, and zoos should embrace them as a valuable management aid.

In view of these assertions, it is interesting to note that Paul Fritz (personal communication), after years of work with chimpanzee rehabilitation at The Primate Foundation of Arizona, has concluded that former circus chimpanzees are more likely to breed than their zoo or lab counterparts. Regarding fatigue from overtraining, Morris has written:

> Far from being overworked, exhausted chimpanzees, these demonstration chimps are by far the healthiest, most intelligent, and most alert that I have ever seen in captivity. They obviously benefit tremendously from their varied and complicated activities and one is immediately struck by the need for introducing some similar kind of occupational therapy for adult chimpanzees and for other primates.

Throughout the world, a variety of habitats have been created for the purpose of ape-keeping. Although my personal observations of European zoos are limited, I am indebted to my former student Susan Wilson who recently visited many European ape facilities and shared her findings with me. It is a tribute to the importance of captive breeding programs that Ms. Wilson was awarded a Thomas J. Watson Fellowship in order to evaluate European ape-keeping efforts.

LABORATORY FACILITIES

Standard laboratory squeeze cages are completely inadequate for the keeping of apes. For clinical treatment or for invasive research efforts they may be required, but their detrimental effect on the behavior of the animals is clear. In fact, in many cases, isolation of an animal which is in need of care can actually contribute to its demise from the combined effects of its ailment and stress. Consider Köhler's (1925) early recognition of the problem of isolation:

> It is hardly an exaggeration to say that a chimpanzee kept in solitude is not a real chimpanzee at all. That certain special characteristic qualites of this species of animal only appear when they are in a group, is simply because the behaviour of his comrades constitutes for each individual the only adequate incentive for bringing about a great variety of essential forms of behaviour.

Clearly, the biggest problems with laboratory housing are the problems of physical size, and social isolation. Moreover, Reynolds (1967) has differentiated between the "hygiene" and "natural" schools of ape-keeping, the former of which predominates among laboratory workers. The benefits of a hygienic habitat are obvious, but the stark cement and steel enclosures are generally not conducive to social activity. This is not to say that cement and steel per se are to blame. A spatious and creative cement and steel enclosure could be both hygienic and natural, but few laboratories have endeavored to build such facilities. It is encouraging to note, however, that agencies of the United States government recently agreed to develop a facility for chimpanzee rehabilitation and captive propagation. Some of the suggested designs for such facilities included innovative concepts such as spatious and well-equipped exercise* areas adjacent to home cages. The laboratory worker must contend with the demands of ready access, restraint, and control of the subject and still meet its psycho-social and physical needs. A ready compro-

*For young orangs, Harrison (1962) has argued:

The next step—to lead a growing baby from the toddler stage into adolescence—is a much harder one to take in a zoo. For this is the time when the young Orang should be taught to move, to exercise his body. He will only do it, though, if you give him something to arouse his curiosity, if he has an incentive and proper facilities. Tables, chairs, and "tea-party" manners are of no interest to him whatever, and you cannot really teach him football or other such exacting games. (p. 156)

Table 7-2. Recommended depth and width (in cm) of orang-utan enclosures. (Adapted from Reuther, 1966)

Zoo	Actual Depth	Recommended Depth (min.)	Actual Width	Recommended Width (min.)	Comments
Dudley (GB)	170		460		wet moat and hot wire
NY Bronx	270	270	430		water level 60–150 cm.
Toronto Metro	330	370	370	370	
Phoenix	370	370	380	380	
San Francisco	370		490		
Twycross (GB)	370	240–270	370		animals have reached up to 210 cm.
Antwerp (Belg.)			400		wet moat, wire mesh barrier
Kansas City	430	370–430	430	430	
Miami Goulds	460		460		water level 30 cm.
Baton Rouge			240–1520	240	wet moat around island

mise has rarely been achieved, but it is an achievement which is highly desirable and not without promise of success.

ORANGS IN ZOOS

The problems of the zoo habitat are quite similar to those of the laboratory. However, some of these common problems are even more difficult to manage in the zoo setting. For example, the health of captive apes is difficult to protect when the public is permitted close proximity to them. The modern solution to this pressing problem has been the use of glass-fronted enclosures. When the surface is strong and clean, a satisfying view of the animals is created, and they are relatively well-isolated from human disease. The glass front also serves to protect the public from flying fecal projectiles, spitting, and water throwing, which are common features of captive ape behavior. A drawback of windows is that they seem to encourage the reduction of barrier distance so that the public may get a close-up view of the animals. Chimpanzees, for example, often respond to these public invasions by kicking and hitting the glass, which encourages public teasing. Glass-fronted enclosures therefore may inadvertently violate flight distance requirements.

Within the zoo enclosure, whatever barriers may be constructed, there are a variety of techniques which may be utilized to enrich the surroundings. To replicate the function of trees and natural cover, steel and concrete can be employed in creative ways. A good example of this is the Hamadryas baboon enclosure at the *Madrid Zoo*. In this setting, the complexity and utility of the structures is striking. Equally functional is the large wooden apparatus which was recently erected at the *San Diego Zoo* for orang-utans, and the complicated, but smaller, version which has been standing at the *Phoenix* (Arizona) *Zoo* for several years. Although subject to wear, wood lends a natural appearance to any enclosure. Even in the construction of human playgrounds, wood structures have become very popular. In the construction of wooden apparatus, volunteers from the community may be employed as has been accomplished by Lee White at the *San Francisco Zoo*.

In her valuable report, White (1978) described her efforts to remodel the orang-utan enclosure at the San Francisco Zoo. With citizen cooperation, at a cost of $1,000, the enclosure was modified in four weeks to resemble an innovative children's playground. The former enclosure (32 feet in diameter) contained only one dead tree with a hanging ladder, tire, and sitting platform. As White explained:

> . . . we decided to add a climbing apparatus utilizing natural materials, preferring pliancy over rigidity and offering a diversity of possibilities for exploration and activity. We sought to provide shade for the animals as well. Loose "toys," all heavy duty and non-toxic, were also to be introduced. Periodical presentation of novel items was also part of the plan (p. 179)

Prior to remodeling the enclosure, the two orang inhabitants (seven and ten years of age) spent approximately 30% of their time concealed under gunny sacks. Play was rare and little climbing/hanging was observed. Although the animals were initially cautious (if not afraid), they soon began to explore the new habitat, exhibiting behaviors which never occurred while in the previous enclosure. White reported that the new habitat affected the pair's status relationship, resulting in greater assertiveness by the younger male, diminished pacing by the female, decreased time under sacks, and increased "arboreal" movement. Clearly, as this study demonstrates, an enriched habitat benefits both inhabitants and viewers. A useful manual for constructing such structures has been written by Hewes (1974). At the *Zurich Zoo* in Switzerland, bamboo has been utilized to construct arboreal pathways* for a variety of species. Another way to use bamboo is as a background, where it can be separated from the animals by wire or glass. Although an illusion, the use of growing foliage as background creates a superior educational display, and the growth may be culled periodically to obtain browse.

A subtle dimension to captive habitat design is that of color. Whether a colorful background is important to the apes is unknown, but the public's perception of these surroundings can certainly be improved by attention to color. It should be admitted, however, that cosmetic improvements are no substitute for enlightened design. Furthermore, it is clear that esthetic conceptions can often obscure functional considerations. This is often the case when well-known designers are employed to create *works* of art, rather than *working* art.**

Sheldon Campbell has discussed this as follows:

> Zoo designers are not as a rule animal behaviorists. The disaster that can be wrought when an otherwise outstanding architect attempts to design a zoo was well illustrated by what happened in Los Angeles . . . Exhibits designed for the public but not for the animals posed many problems . . . Fortunately, the original mistakes in design are being rectified by the zoo's present director, Dr. Warren Thomas, a man firm in his conviction that zoo exhibits should be pleasing to the public, practical for the staff, and natural for the animals. And to those three criteria virtually every zoo director in the world would add in chorus "and economical to build." (pp. 73-74).

---

*The application of these principles to zoo design may come to be known as "apescaping."

**Harrisson has discussed the need for aboreal considerations in her 1962 book as follows:

Orangs live in trees. Their strong arms are made to swing from swaying branches; their feet to grip and climb, not to walk on the ground. Yet they generally get the same sort of facilities and treatment in zoos as *terrestrial* Gorillas and Chimpanzees. Often they are kept in one cage with the latter—to the overwhelming disadvantage of the Orangs. To a certain extent young Orangs adapt themselves to walking on concrete. But once they get older they usually grow resigned to sitting on an upper shelf (if there is one). By that time the rough surface of the floor, or wrong dirt, or disease, have rendered their long, shiny hair dull and sparse, moth-eaten and short. Some Orangs appear entirely 'naked' and their fur never grows again. In the wild . . . such pitiful creatures could not occur. (p. 134–135)

ANIMAL PARKS

The concept of outdoor, seminatural animal parks has flourished in the last decade. Because these parks allow the animals a considerable amount of space, there are a number of difficulties where apes are concerned. Because of the apes' arboreality, fences must be equipped with sheet metal topping to prevent escape. Because of their great strength and manipulative abilities, the fence itself must be exceedingly well constructed.* Due to the cost of such alterations, "new zoo" designs for apes have favored islands surrounded by water or dry moats. Often, moated, outdoor exercise areas adjoin indoor housing, particularly in cities where the climate is not conducive to ape-keeping.

The more economical solution of ape-keeping is the use of natural or man-made islands surrounded by water. In all cases, however, the island *must* be equipped with protective housing, to provide both cover and heat. Moreover, these areas must be secure in order to facilitate capture. This is best accomplished when the animals are locked up at least once each day for feeding and close-up inspection. While the rain and cold are an ever-present danger to health, equally hazardous is the water barrier. Opinions vary as to the reasons for drownings, but few apes survive a fall into water, since they are apparently not capable of swimming.

For five years (1972-1977), the Yerkes Regional Primate Research Center maintained chimpanzees on *Ossabaw Island*, off the coast of Georgia. Despite an optimistic report by Wilson and Elicker (1976), the animals were removed after a number of them had died. The danger of drowning, and the difficulty of maintaining the animals under absentee management conditions contributed to the decision

---

*The ability of orangs to destroy their surroundings is well known to students of the species:

> Unhappily there is one ready excuse for not giving facilities for exercise and play; especially not to Orangs. They have one ability which distinguishes them from the other apes: they take everything to pieces, an attitude also referred to as their 'engineering skill . . .' The more Orangs are encouraged by kind keepers who give and replace things, the more ingenious they get in destroying them again. After some time everybody—directors as well as keepers—give up; the cages become bare, no more ropes, except perhaps a short strong bit of plastic or an iron chain; no sacks; one huge strong beam instead of a series; one polished shelf (harder to tackle) instead of a rough one; and a metal swing, if any. But this kind of equipment is of no abiding interest to the curious Orang. Once tested and explored, moving about in it becomes a matter of routine–like the endless walkabouts of a tiger behind the iron bars of his cage. (Harrisson, 1962, p. 142)

As Brambell (1975) has also pointed out, heavy males will stop climbing if they distrust the strength of a structure. Thus the materials of a climbing structure must withstand the rigors of the user.

to remove them. Obviously, it is better to utilize islands which can be easily managed on a daily basis. Where a large natural island can be properly managed, Wilson and Elicker mentioned the following advantages:

a. Physical health can be easily monitored and any necessary medical treatment can be administered;
b. Intensive and extensive behavioral observations should be less difficult than in the wild state;
c. The maintenance cost per individual is less than in the captive state, as is the cost of building the initial facilities;
d. Reproductive rates are at least as high as those observed in captive settings;
e. The infants produced are psychologically and behaviorally more normal; and
f. Previously experimental, behaviorally abnormal adults will experience some degree of rehabilitation towards the more normal species-typical behavior observed in wild populations.

The island concept has also been discussed by Reynolds (1967) in the advocacy of what he has called *apelands*. Like the commercially popular *marinelands*, these apelands, as Reynolds sees it, would provide for public exhibition and entertainment. When not in the public view, the animals would be allowed free access to a small forest* with perimeter feeding stations where animals could be checked and/or captured. Daily observation by wardens would be a regular maintenance task, while lab work would be conducted in adjoining buildings. Those animals which were used in research would be periodically shuffled from forest to lab and back again. In suggesting the necessary elements for a successful enclosure, Reynolds listed the following:

---

*Harrison has observed that captive orangs delight in access to trees and browse:

This Orang demonstrated, wildly enough, how he wanted to live. Then *why not build him an enclosure round a tree?* And if this is impossible, and I see no reason why it should be, why not give him at least some large branches to construct a nest within his cage? . . . What is required for these growing youngsters are grounds with trees where they are left to do as they please, preferably with a number of teen-age playmates of their own kind. Ideally the grounds should be large with many trees in them. But a small area with only one tree is better than to leave the animals in the cage at all times. There is no need to fear that they will forget in the long winter months of a temperate climate what they can do in a tree. (1962, pp. 157–158)

. . . trees, a grassy paddock, moat with electric wire, cover (rocks) and dividers, rain shelters, fresh-cut branches for nest construction, interspecies housing (where space allows it), indoor housing with climbing materials, free access to indoor and outdoor facilities, ad lib food, and access to conspecifics.

The electric wire mentioned by Reynolds is one solution to the drowning danger posed by water. With this technique the animals are discouraged from approaching the water. At *Kingdoms Three Animal Park* in Atlanta, our research team constructed a similar device in order to discourage activity in the water. Our barrier was a chain supported by pipes which rings the island, giving the animal an opportunity to grab onto something should it slip into the water. We have had no incidents since the installation of the chain, despite the fact that two apes had drowned at the park in previous years. Van den Bergh (1959) also utilized a barrier system to prevent drownings as follows:

. . . We have safeguarded ourselves against a similar occurrence (drowning) by placing a wire mesh barrier between the shallow area of water near the enclosure and the deeper section of the moat. Since its installation, the barrier has twice saved two very lively young gorillas which fell into the water and were able to get out again on their own accord.

Another problem with islands is that unless they are quite large, the inhabitants will soon eat their way out of house and home. Therefore, it is often necessary to build artificial structures which provide shade and refuge should foliage be depleted.

Where the weather is suitable for island living (tropical or subtropical settings are to be preferred), their advantages are obvious. Islands can be an inexpensive, natural alternative to the elaborate structures which have become the norm in modern zoo design.

HABITAT FUNCTIONS

Besides the primary functions of feeding, care, and capture, the captive habitat should be constructed so that public viewing and daily observations are unobstructed. The former is complicated by some of the animals' needs that I have previously discussed. The latter can be facilitated by building into the enclosure observation areas, closed-circuit television access points, and/or one-way observation windows. As I have emphasized, a well-equipped research program is an essential feature of successful captive animal management. Clinical judgments can be enhanced when the veterinarian is supplied daily behavioral data, and with regular observation, problems can be objectively verified and valid solutions are more likely to be found. Behavioral research on the apes is a matter of critical importance, since we are obligated to breed these endangered animals in captivity. As we have seen, an understanding of behavioral problems can improve the state of the apes and insure that their captive propagation will be successful.

CONCLUSIONS

Throughout our history of ape-keeping, we have constructed occasionally good, but more often bad and even ugly habitats for these sensitive and intelligent creatures. In reviewing the literature and the lore of ape-keeping, it is clear that we still do not have definitive, empirically derived evidence on the social effects of habitat. Still, there are some testable hypotheses, solid and compelling influences, and many useful observations on which to rely. There is much agreement on optimum habitats, and a general trend toward improvement in design, management, and display.

To aid further progress, we need more research on these applied questions, and greater cooperation in bringing about more positive changes. Great ape breeding loans should be monitored by scientists in order to pinpoint problems and procedures, and there is a pressing need to share information and encourage outside investigators from our universities.

While I have suggested a great deal here, I am painfully aware of the need for more convincing data, and ultimately more compelling arguments. Habitat *can* be assessed as an independent variable which affects behavior. We can count the number of corners, measure area and volume, enumerate manipulanda, and record activity. We can conduct pretests and post-tests (even protests), and we can combine our notes to increase our "n." Hence, we *can* do research on the effects of captivity. Indeed, we *must* do it. As Yerkes and Yerkes (1929) so neatly expressed it:

In the past, knowledge of anthropoid life has grown haltingly, irregularly, uncertainly, because of fragmentary, unverified, and often unverifiable observations. Because of adverse conditions, investigations have been relatively unsatisfactory: witness, attempted contributions to knowledge of courtship, mating, and other important aspects of the reproductive cycle, of life history, rate and conditions of growth, mode of life, heredity and acquired modes of response . . . Obviously, many things have been done poorly, though at great pains, which under carefully planned and appropriate conditions might have been done well (p. 590).

Now is the time to rectify this situation and to initiate systematic, cooperative inquiries into the state of the ape in captivity. The challenge is an urgent one if we are to successfully ensure that these endangered anthropoids are to survive the relentless growth of human civilization. As Robert Sommer (1974) has concluded:

If living creatures cannot be left in their original habitat, the least that can be done is to place them in natural and responsive surroundings—natural so that their character is not warped, and responsive so that their individuality and creativity are firmly respected. (p. 69)

Figure 7-1. The many faces of captivity (continued). (E. Zucker photos)

Figure 7-1. The many faces of captivity. (continued) (E. Zucker photos.)

8 Conservation of the Orang-utan

> Every field observation, apart from scientific value, in itself may assist materially in developing better ways of keeping Orangs in captivity and conserving them in the world. At the moment, whether we like it or not, we have to face the fact that it looks like somewhere near a forty-sixty chance of 'the captive solution' being the only way to keep these great apes alive for our great grandchildren—at least.
>
> (Harrisson, 1962, p. 173)

There are three main conservation problems common to Borneo and Sumatra: human exploitation, human population expansion, and deforestation. Each of these categories is composed of traditional and more recent elements, and each is related to the other.

Historically, as we have seen, orang-utans have been exploited by human beings, first as a source of food, later as a substitute for other humans when cannibalism was suppressed, and more recently in the pet and zoo trade (cf. Harrisson, 1962). In both Malaysia and Indonesia, people are known to keep orang-utans in their homes. This problem is particularly acute among logging personnel where many young orangs have been found, having been captured during the course of deforestation.

In the second category, the islands of Borneo and Sumatra have suffered the same large population increases as have other rapidly developing countries. As the demands of human agriculture have increased, orang habitats have shown a concomitant decrease. In this sense, orang-utans and people are in direct competition for scarce resources, e.g., space and food.

Finally, deforestation, as we have seen, is affected by human population expansion. Both the logging industry and the pressures of traditional agriculture lead to the removal of native trees, resulting in barren space. When people move in to cut down trees, both men and machines contribute to the departure of orang-utans as well as other animals, which must move deeper into the dwindling forests. With food sources already scarce and dispersed, the remaining animals must compete for the remaining resources. As they are crowded closer together, conflicts may lead to further attrition.

These are serious problems. As we consider them in this chapter, we will review the suggestions of field workers and conservationists. Hopefully, from this collection of ideas, some workable solutions can be found. It will be tragic indeed if orang-utans fail to survive these many pressures.

Historically, the greatest danger to orang-utans has been from human hunting or capture. Orang remains have been located in limestone caves throughout Asia, suggesting that prehistoric humans hunted orangs for meat or for their skulls as "trophies." If we assume, as has Rijksen, that the early peoples of Southeast Asia were primarily fruit eaters, it is likely that humans and orang-utans were in direct competition for scarce resources. In more recent times, orangs have been captured to serve as pets, or to be transported to zoos and laboratories. While predation by other animals, such as the tiger and clouded leopard, has doubtless accounted for some losses in earlier times, predation by other animals is insignificant when compared to human pressures.

One of the most tragic features of the illegal orang-utan traffic is that so many animals must die in order to capture one infant for sale. Barbara Harrisson (1961) discussed this problem in the following passage:

> . . . this method of collection not only involves killing *one* mother for each live baby Orang–who incidentally remains during the first few years of his orphaned life a highly delicate creature, especially prone to all sorts of human infections. It is at best an underestimate–under present conditions of trade and travel in this part of the world–to allow for *three deaths for each live baby leaving the habitat countries* to go to a zoo or dealer abroad, namely: two mothers shot dead and one baby dying "in transit"–not to mention the number of the yet unborn killed by the killing of mothers! (p. 242)

To further amplify this point, Harrisson pointed out that the last zoo collector permitted to operate *legally* in Sarawak in 1946 removed *fourteen* or more live orang-utans. She asserted with certainty that *twenty* others were known to have died as an immediate consequence (p. 167). Of course, modern zoo workers are aware of this unsavory history and have dedicated themselves to the support of con-

Figure 8-1. Young orang-utans at Sumatran rehabilitation center (Photo courtesy of H. D. Rijksen).

servation. In fact, with recent breeding successes, if habitats were available, we could soon begin to return some orang-utans to the wild.

Summarizing her concern for the loss of both mothers and infants through capture Harrisson expressed her views as follows:

An Orang mother gives birth to a single offspring only, about five times during her normal lifespan; and infant mortality is probably high in the wild, also. Development to maturity is slow (ten to thirteen years) and the infant's dependence on mother's and group education lasts several years. It seems therefore reasonable to conclude that a *small* Orang group–possibly also in disadvantage through former killings of mother with a much stronger element of old and adult males–is just about capable of reproducing its own number under the present conditions and that *each killing or trapping results in an equal reduction of the remaining habitat population.* (1961, p. 243)

EARLY RECOMMENDATIONS

The following conservation recommendations have been extracted from a larger list as suggested by Carpenter in 1937. These recommendations are especially interesting because they represent views which have only recently become law.

VII. It is believed that in order adequately to protect the orang-utan, the conservation program must extend beyond that of mere protection to cultivation. This is necessary because much of the most suitable habitat for orang-utans has been destroyed. It is believed that this situation has resulted in a decreasing population of this important anthropoid.

VIII. As a result of the importance of the orang-utan as a zoological organism, and since an urgent need exists for reliable and adequate data on the population and changes which are occurring in this animal type, it is strongly recommended that the *killing* and *capture of this animal for trading or exhibition purposes be completely stopped* and that its use for accredited but limited scientific purposes alone be permitted.

IX. On the basis of the improbability that orang-utans will attack man, and even if such were to occur in very rare instances, klawangs* would be adequate protection from the rather slow animal, it is recommended that the soldiers in Atjeh be forbidden to shoot orang-utans under any circumstances.

X. It is suggested that when large tracts of land are given in concessions and when these areas include habitats of the orang-utan, the apes be transferred to primary forest either by driving them or by capturing and transporting them, rather than capturing them for sale or destroying them. In situations where it seems more feasible and desirable, plots of forest could be left standing for the specific purpose of harboring both orang-utans and gibbons.

XI. It seems very necessary and altogether desirable that the Indies- and Netherlands Committees which are actively interested in and responsible for wild life protection, engage in an educational compaign in order to instruct natives, soldiers, naturalists and others in the interesting natural history of Atjeh. It is specifically recommended that all military posts and all officers receive regularly the publications on the many forms of Atjehnese plant and animal life and that lectures be arranged for the officers when they have conventions in Koeta Radja.

These recommendations are the same recommendations that we would give today and, in fact, it is now the policy of the Malaysian

*clubs

and Indonesian governments to prevent illegal export, prohibit killing, and to promote conservation education with respect to resident orang-utans. Where much-needed work is required is in the preservation of indigenous forest habitat and the establishment of fully protected regions for orang-utan habitation. A major problem continues to be *enforcement* of existing law, and a pressing need to increase public and corporate *awareness*.

In her 1961 book, Barbara Harrisson also compiled a list of recommendations for the conservation of orang-utans. I list them here as follows:

1. The burden for export *control* must not be left simply to the countries which have wild Orangs—namely the Republic of Indonesia, Sarawak and North Borneo.
2. *Much higher standards* must also be set for the care of apes in captivity, with *particular reference to breeding*. Standards should be agreed upon internationally and with the relative governments informed at a high level.
3. The zoos must exercise more *self-control and less selfishness* in acquiring ape exhibits, especially in acquiring "at all costs" for their own commercial reasons, single specimens of Orangs . . . which cannot possible even be happy, let alone breed.
4. A specialist conference should be called to discuss measures to promote *breeding in captivity* and an international research foundation be set up with this immediate goal: the pooling of all information on breeding in captivity and the developing of further breeding and conservating techniques (in cooperation with responsible zoological societies).
5. That, as a corollary of the above, an international research project be sponsored to assist the *Indonesian and other governments*, both in ascertaining present Orang population accurately, and in devising practicable methods to preserve their status.
6. That dealings in protected animals (including the Orang) by private businessmen (i.e. animal dealers) be outlawed. Zoos and scientific institutions should acquire these animals through a recognized international organization on a nonprofit basis.
7. . . . form at least one proper sanctuary for Orangs before it is too late; and to enlarge the scope and personnel of Orang research, with outside assistance, to study those other urgent problems which can only be seen *in the wild*. (p. 172)

RELOCATION TO THE WILD

Barbara Harrisson's work in the early 1960's was the first effort to solve the problem of abandoned orang-utans. Many animals, orphaned as a result of their acquisition for research, zoos, or the pet trade, were left over when the game laws of Malaysia and Indonesia were strengthened. Harrisson began her work at Bako in Sarawak and it was ultimately expanded by the Sabah Forest Department at Sepilok.

REHABILITATION TO THE WILD

The rehabilitation of young orang-utans was originally composed of the following basic three steps (Harrisson, 1962):

I. *Education to semi-wilderness in garden areas of Kuching.* Behavior response of that first phase . . . : a.) Learning in trees, climbing possibilities/incentives; b.) Emotional adjustments, companions, sounds.

II. *Transfer to a jungle enclosure,* prior to release . . . includes: c.) Response to jungle surroundings; d.) Readjustment of human relationships.

III. *Release to freedom* . . . includes: e.) Learning to wander in trees; f.) Relations with companions/sounds.

This early success led to more recent efforts at Ketambe and Bohorok in Sumatra and Tanjong Puting in Kalimantan. The subject of considerable publicity, the success of these rehabilitation/relocation programs was recently assessed by the field worker John MacKinnon (1977) who had studied orang-utans in both Borneo and Sumatra (cf. MacKinnon, 1974; 1976). As MacKinnon argued:

> . . . far from being underpopulated orang-utans are more frequently overcrowded because their forest habitat has been destroyed faster than their numbers have been depleted; homeless orang-utans become bad tempered nomads, seriously disrupting the social harmony and breeding success of surviving populations. There are too many orang-utans for the remaining forests, not too few. (1977, p. 698)

A further complication mentioned by MacKinnon is that rehabilitants released into the wild may transmit human infectious diseases to the wild population. Orang-utans are susceptible to malaria, nematode parasites, tuberculosis, Herpes and other viral infections and may have little natural immunity to such disorders. Even more alarming, according to MacKinnon:

> Orang-utans have sometimes been used for sexual purposes in Bornean long houses and the possibility of a rehabilitated animal spreading a venereal disease to wild animals cannot be ignored. The return of captive animals to healthy wild populations can no longer be justified. (1977, p. 698)

Another problem with rehabilitation schemes is the difficulty of training a human-dependent, terrestrially adapted, gregarious animal to be arboreal and afraid of human contact. In the transitional stage of training, rehabilitants are subject to easy predation by clouded leopards and the like. Moreover, it is exceedingly difficult to ever determine whether a rehabilitant has been successfully established in the wild, once it has penetrated the jungle. In an earlier discussion of nursery rearing I suggested that the existence of nurseries may compel staff to remove infants from their mothers prematurely. In a similar manner, MacKinnon has been concerned about the compelling potential of rehabilitation stations:

> The stations at Bohorok and Sepilok have been promoted as tourist attractions, which probably makes it even harder for the young apes to forget the human world they have discovered and revert to the wild life they hardly know. Indeed so much money has been invested on developing the Sepilok station that one can envisage the Forest Department having to procure new orang-utan inmates to exhibit to tourists long after the need for rehabilitation in Sabah has ceased. (1977, p. 698)

Given these clear problems, what then is the value of rehabilitation stations, if any? One important function is that these stations serve as visible reminders that the governments and concerned international organizations will not permit the illegal exploitation of these animals. Moreover, as way-stations for formerly caged animals, the rehabilitation station is necessary. MacKinnon also believes that the procedures which have been worked out by dedicated rehabilitation workers can be adapted to relocation projects in the face of increasing losses of habitat. In the long run, rehabilitation is of little use in the protection of the genus *Pongo*. Conservationists now define the problem in larger terms. The orang-utan can only be saved if the entire forest ecosystem is saved. To do so, as in all successful conservation efforts, will require massive education of the indigenous peoples. For if the people do not want the forest saved, it cannot be saved. The potential economic value of forest preservation must be stressed so that the people realize that conservation also makes good economic sense. It is in this complex educational effort that the rehabilitation centers can be especially helpful. As MacKinnon so cogently concluded:

> The rehabilitation centres must be flexible enough to adapt to current problems if they are to be effective and command our respect and financial support. They must avoid the trap of becoming labelled as expensive places where soft-hearted women fondle furry substitute ape children. The centres should really call themselves conservation field stations, for not only is rehabilitation no longer their main function but also the orang-utan itself is only a familiar and personal figurehead for the whole forest ecosystem, and it is the whole system that must be conserved. (1977, p. 699)

At the Sixth Congress of the International Primatological Society, a roundtable discussion on rehabilitation and conservation was held and later summarized in a paper by Borner and Gittens (1978). It was generally agreed that, as MacKinnon (1977) had earlier suggested, it is "irresponsible" to release former captives without knowing the effect it may have on the area's resident population, especially in view of the possibility of spreading disease. Thus, plans were revealed to release rehabilitants into protected areas where no wild orangs are currently to be found. Given the efforts to reintroduce orangs, Borner and Bittens nonetheless concluded;

> It is obvious that one cannot save this population by releasing some thirty animals. Today orang-utans are not primarily endangered by poaching, but by the destruction of their habitat caused by logging, oil exploitation and agriculture activities. The most important task of a rehabilitation station must be propaganda and conservation education (p. 102)

By doing so, the participants agreed, the enactment of conservation laws will be facilitated.

Orang-utan reintroduction was considered to be especially suitable in that it is not necessary to introduce them into an existing group. Thus the relatively solitary nature of orangs makes their management somewhat easier than it is for gorillas and chimpanzees.

The roundtable group also concluded that while population resettlement should be attempted, a better method might be to lightly log the forest in slow rotation. This technique would permit the resident animals to voluntarily move from logging areas to locations undergoing successful regeneration.

Protected islands (cf. also Chapter 7) where semi-wild breeding colonies could be established was also suggested as a means to ensuring the continuation of the species when, at some future date, suitable areas in their natural range became available.

If Thorington's (1978) "glum view" of the future comes to pass, there will be little or no tropical forest left in the world. In his view, there are three possible outcomes, each more optimistic but less probable than the former:

1. All unprotected tropical forests will be destroyed, and most wild primates will disappear.
2. There will be protected islands of tropical forest, including newly established areas not currently protected, whereby most primates and other fauna will be preserved.
3. Due to human disasters, population increases will be slowed, and the rainforests and primates will be at least as well off, or better off, than they are today.

Because the latter possibility is hardly worth wishing for, the future of nonhuman primates, according to Thorington, rests with the preservation of adequate habitats and unconventional developments in technology, e.g. "a more saleable veneer than that derived from tropical hardwoods," or the discovery of "more efficient ways to extract protein from a tropical forest than to cut it down, plant grass, and raise cattle (p. 4)."

When one carefully considers the conservation of the orang, an irony emerges. As we fight the battle to promote its survival in the wild and in captivity, we discover that it is the quality and quantity of habitats that will determine our ultimate success in both settings. In one, it is a problem of adequate food, space, and safety. In the other, it is primarily a matter of space, complexity, and social experience. And so it is that even as we address the conservation issue, the field and captive settings are interrelated. We must work together, and solutions must be developed by those who understand the problems of both captivity and wilderness. By preserving the latter and enriching the former, orangs will survive and we may one day be required to reckon with the overpopulation of *Pongo*. It will be a welcome challenge in view of current prospects of this vanishing species. Let there always be orang-utans.

Epilogue: Does the Orang-utan have a future?

> Scientific observations on Orangs have been carried out by a very few zoos . . . Unfortunately much of this work was destroyed as a result of the Second World War. It is now up to those fortunate enough to house Orangs capable of breeding—to carry on that work and to bring up animals in zoos that will be capable of breeding a second and third generation.
>
> (Harrisson, 1962, p. 155)

In this book I have reviewed laboratory, zoo, and field studies on the behavior of orang-utans. The literature clearly indicates that this red ape is unique among the family members. It is more solitary than gorilla or chimpanzee, and it is highly arboreal unlike the other two. While its "personality" may be stoic, it is nonetheless capable of maintaining intense social relationships. The mother-infant bond is particularly rich and idiosyncratic. Its cognitive abilities are well renowned, but incompletely understood. A determined technician, the orang in captivity is always a threat to escape. Slow moving, were the ape to make an escape, it would likely be unhurried. We have learned, too, that the orang-utan is a selective feeder in the wild, ranging widely in order to locate its preferred foods. In captivity, the orang adjusts well, despite the prevalence of habitats which lack imagination and proper space. The red ape is in danger of extinction. What can we do about it?

Research is a first step. Only in the past decade have we begun to understand the nature of this creature, despite the fact that orangs were common in many world zoos.

In 1929, the Yerkes' expressed their dismay at the paucity of research on apes in captivity in the following paragraph:

> For decades the zoological gardens of the world have held captive specimens of gibbon, siamang, orang-outan, chimpanzee, and gorilla. Often the individuals have lived for many years, and occasionally they have bred, in reasonably satisfactory environment. Yet, almost without exception, the scientific use of these exhibition specimens has been neglected. Evidently there is opportunity for some progressive zoo-

logical-garden director to lead the way and establish a fashion by converting his establishment into a center for biological research without undesirably sacrificing its primary function of entertainment and education. (p. 582–583)

In her 1962 book Harrisson also addressed this point:

Few zoos have taken the trouble to observe their animals regularly and publish results. If only they had, we might have a complete picture of the behaviour of our nearest relatives—at least, in captivity, where so many have suffered and died to delight millions of civilised humans. (p. 154)

By studying wild orang-utans we can learn about their requirements in nature, and act to properly manage their habitats. Without the necessary information, legislation cannot be produced. An understanding of the free-ranging animal may also lead to improved cage designs and management policies for those which are legally held captive. We must do what we can to insure that captive orangs reproduce and raise their young, and the remaining wild populations must be protected from human exploitation. As we have seen, the acquisition of orangs for public and personal amusement has tragically diminished their numbers.

Human exploitation of orang-utans even contributes to the confounding of field studies, as our knowledge of orangs comes increasingly from easier-to-observe "rehabilitants." Moreover, it may be that some observations have been made on escaped or released "captives," the behavior of which has distorted some of the researchers' conclusions. We know, for example, that monkeys and apes living under conditions of restriction will develop *deprivation acts* (cf. Berkson, 1968) such as self-biting, rocking, and eye-poking. That at least two of Davenport's (1967) wild subjects may have been former captives is suggested by his following observation:

A curious behavior was noted in the case of two animals for which I offer no explanation. Both of these animals engaged in what might be called "saluting," that is, holding an open hand up to the side of the face, the forefinger proximal to the outer edge of the eye. This position was held from 20 to 40 sec. There was neither rain nor bright sun from which to shield the eyes. (p. 253)

From the work of Berkson (1968), Mitchell (1968), and Erwin et al. (1973), it is clear that "saluting" (or "eyepoking" as it is also manifest) is a *deprivation act* of primates that have lived in captivity.

Figure 9-1. The future of the orang-utan is at best uncertain (T. Maple photo).

Thus, support for scientific research, both in field and zoo, wildlife advocacy through international conservation agencies such as the World Wildlife Fund, lobbying and legislating for improved captive habitats, and a personal committment to speak up on behalf of the orang-utan is the least that we can do to ensure their survival. Does the orang-utan have a future? Its future is our future. The decisions which we make will reflect our resolve to preserve the rich fauna and flora of this planet. If this large, clever creature should be driven to extinction, can others be far behind?

I have been privileged to study the orang-utan. I hope that this book will make a difference. If the reader has been moved to contribute to their welfare, to learn more, or simply to better appreciate them, the effort has been worthwhile. However ignorant we may feel, we have learned much about orang-utans. We certainly know better than to lose them.

References

Abel, C. 1818. Narrative of a Journey in the Interior of China, and of a Voyage to and from that Country, in the Years 1816 and 1817. Longman and Co., London (1818). Cited in Yerkes and Yerkes, *The Great Apes*, 1929.

Altmann, S. A. 1967. The structure of primate social communication. In *Social Communication Among Primates,* S. A. Altmann, ed., Chicago, University of Chicago Press, 325–362.

Asano, M. 1967. A note on the birth and rearing of an orang-utan at Tama Zoo, Tokyo. *International Zoo Yearbook*, **7**, 95–96.

Aulman, G. 1932. Geglückte Nachzucht eines Orang-utan im Dusseldorfer Zoo. *Zool. Gart.* (*Leipz.*), **5**, 81–90.

Barash, D. 1977. *Sociobiology and Behavior*, Amsterdam, Elsevier.

Bauman, J. E. 1923. The strength of the chimpanzee and orang. *Sci. Month.*, **16**, 432–439.

Bauman. J.E. 1926. Observations on the strength of the chimpanzee and its implications. *J. Mammal.*, **7**, 1–9.

Beach, F.A. 1976. Sexual attractivity, proceptivity, and receptivity in female mammals. *Hormones and Behavior*, **7**, 105–138.

Bemmel, A. C. van. 1959. Keeping apes at Rotterdam Zoo. *International Zoo Yearbook*, **1**, 16–18.

Bergh, W. Van den. 1959 The new ape house at the Antwerp Zoo. *International Zoo Yearbook*, **1**, 7–12.

Berkson, G. 1968. Development of abnormal stereotyped motor behaviors. *Dev. Psychobiol.*, **1**, 2, 118–132.

Berkson, G., Mason, W. A., and Saxon, S. V. 1963. Situations and stimulus effects on stereotyped behaviours of chimpanzees. *J. Comp. Physio. Psych.*, **56**, 786–792.

Bernstein, I.S. 1969. A comparison of the nesting patterns among the three great apes. In Bourne, G. H. (ed.), *The Chimpanzee*, Basel, Switzerland, S. Karger, 393–403.

Bingham, H. C. 1927. Parental play of chimpanzees. *J. Mammal.*, **8**, 77–89.

Bolles, R. C. 1975. *Learning Theory*. New York, Holt, Rinehart, Winston.

Bolwig, N. 1959. A study of the nests built by mountain gorillas and chimpanzees. *South African J. of Science, 55,*

Bolwig, N. 1959. A study of the behaviour of the Chacma baboon. *Behaviour*, **14**, 136–163.

Borner, M. and Gittens, P. 1978. Round-table discussion on rehabilitation. In Chivers, D. J. and Lane-Petter, W., Eds., *Recent Advances in Primatology, Volume II*. New York, Academic Press, pp. 101–105.

Boulle, P. 1963. *Planet of the Apes*. New York, Vanguard.

Bourne, G. H. 1971: Nutrition and Diet of Chimpanzees. In Bourne, G. H. (ed.), *The Chimpanzee*, vol. 4, 373–399.

Bourne, G. H. and Cohen, M. 1975. *The Gentle Giants*. New York, G. P. Putnam's Sons.

Brambell, M. R. 1975. Breeding orang-utans. In R. D. Martin (ed.), *Breeding Endangered Species in Captivity*, London, Academic Press, pp. 235–243.

Brookfield Bison, 1973. Hand raising Hahna . . . to be an orang. **8**, 4, 1.

Campbell, S. 1978. *Lifeboats to Ararat*. N.Y., Times Books.

Carpenter, C. R. 1938. *A survey of wild life conditions in Atjeh, North Sumatra: with special reference to the Orang-utan*. Communication No. 12 of the Netherlands Committee for International Nature Protection, 34 pp.

Chaffee, P. S. 1967. A note on the breeding of Orang utans at Fresno Zoo. *International Zoo Yearbook*, **7**, 94–95.

Chamove, A. S. 1978. Therapy of isolate rhesus: different pattern and social behaviors. *Child Development*, **49**, 43–50.

Chevalier-Skolnikoff, S. 1974. Male-female, female-female, and male-male sexual behavior in the stumptail monkey, with special attention to the female orgasm. *Arch. Sex. Behavior*, **3**, 95–116.

Clifton, L. D. 1976. Interspecific social behaviors of infant chimpanzees and infant orang-utans in captivity. Honor's Thesis, Emory University, Atlanta, Georgia.

Coffey, P. Sexual Cyclicity in captive orang utans. *Jersey Wildlife Preservation Trust Twelfth Annual Report*, 1975, pp. 54–55.

Cross, H. A. and Harlow, H. F. 1965. Prolonged and progressive effects of partial isolation on the behavior of macaque monkeys. *J. Exp. Res. in Personality*, **1**, 39–49.

Cummins, M. S. and Suomi, S. J. 1976. Long-term effects of social rehabilitation in rhesus monkeys. *Primates*, **17**, 1, 43–51.

Cuppy, W. 1931. *How to tell your friends from the Apes*. New York, Liveright Publishing Co.

Cuvier, F. 1811. Descriptions of an ourang outang; with observations on its intellectual faculties. Transl. 1810 publ., *Philos. Mag.*, London, **38**, 188–199.

Dann, D. D. 1977. A closer look at man's closest relatives. *The U.C.L.A. Monthly*, **7**, 6, pp. 1–2.

Darwin, C. R. 1872. *Expression of the Emotions in Man and Animals.*

Davenport, R. K., Jr. 1967. The orang-utan in Sabah. *Folia primatologica*, **5**, 247–263.

Davis, R. R. and Markowitz, H. 1978. Orangutan performance on a light-dark reversal discrimination in the zoo. *Primates*, **19**, 4, 755-760.

Deets, A. S. and Harlow, H. F. 1974. Adoption of single and multiple infants by rhesus monkey mothers. *Primates*, **15**, 193–203.

Dennon, M. B. 1977. Affiliation, affection, and attraction in captive adult pairs of orang-utans. Honor's thesis, Emory University, Atlanta, Georgia.

De Silva, G. S. 1972. The birth of an orang-utan at Sepilok Game Reserve. *Int. Zoo Yearbook*, **12**, 104-105.

DeVore, I. (ed.). 1965. *Primate Behavior*. New York, Holt, Rinehart and Winston.

Doorn, C. Van. 1964. Orang-utans. *Freunde des Kölner Zoo*, **7**, 3–9.

Draper, W. A. and Menzel, E. W. 1964. Size and distance of food: cues influencing the choice behavior of orang-utans, pp. 186–190.

Elder, J. H. and Yerkes, R. M. 1936. Chimpanzee births in captivity: a typical case history and report of sixteen births. *Proc. Roy. Soc. Lond.*, **120**, B, 409–421.
Ellis, J. 1975. Orangutan tool use at Oklahoma City Zoo. *The Keeper*, **1**, 5–6.
Elze, K., Zschintsch, A., and Seifert, S. 1971. Zum Schwanger-schaftsnachweis beim Orang-ütan (*Pongo pygmaeus*) durch quantitative chemische ostrogenbestimmung im Urin. Verh. Ber. XIII. Int. Symp. Erkrankungen der Zootiere, Helsinki.
Erwin, J. 1976. Aggressive behavior of captive pigtail macaques: spatial conditions and social controls. *Lab. Primate Newsl.*, **15**, 1–10.
Erwin, J., Maple, T. and Mitchell, G. (eds.). 1979. *Captivity and Behavior: Primates in Breeding Colonies, Laboratories and Zoos*. New York, Van Nostrand Reinhold.
Erwin, J., Mitchell, G., and Maple, T. 1973. Abnormal behavior in non-isolate-reared rhesus monkeys. *Psychol. Rep.*, **33**, 515–523.
Essock, S. M. Disruption of performance of a lab-born orang-utan after introduction of irrelevant foreground cues. *Percept. Motor Skills*, 1975, *40*, 645–646.
Essock, S. M. and Rumbaugh, D. M. 1978. Development and Measurement of Cognitive Capabilities in captive nonhuman primates. In H. Markowitz and V. J. Stevens (eds.), *Behavior of Captive Wild Animals*, Chicago, Nelson-Hall, 161–208.
Eveleigh, J. R. and Hudson, C. F. 1973. Successful fostering of a new born squirrel monkey. *Lab. Prim. Newsl.*, **12**, 13–14.
Fox, H. 1929. The Birth of two anthropoid apes. *J. Mammal.* **10**, 37–51.
Freeman, H. E. and Alcock, J. 1973. Play behaviour of a mixed group of juvenile gorillas and orang-utans. *International Zoo Yearbook*, **13**, 189–194.
Furness, W. H. 1916. Observations on the mentality of chimpanzees and orang-utans. *Proc. Amer. Phil. Soc.*, Phil., **55**: 281–290.
Galdikas, B. M. F. 1979. Orangutan adaptation at Tanjung Puting Reserve: mating and ecology. In Hamburg, D. A. and McCown, E. R., Eds., *The Great Apes*, pp. 195-233.
Galdikas-Brindamour, B. Orangutans, Indonesia's "People of the Forest." *National Geographic*, 1975, **148**, 4, 444–473.
Gewalt, V. W. 1975. Orang-Utans (*Pongo pygmaeus*) als "Seiler." *Z. f. Süngetierkunde*, Bd. **40**, 5, 320–321.
Goodall, J. 1965. Chimpanzees of the Gombe Stream Reserve. In I. DeVore, ed., *Primate Behaviour*, New York, Holt, Rinehart, Winston.
Goodall, J. van Lawick. 1968. The behaviour of free-living chimpanzees of the Gombe Stream Reserve. *Anim. Behav. Monographs*, **1**, 161–311.
Goy, R. W. and Resko, J. A. 1972. Gonadal hormones and behavior of normal and pseudohermaphroditic female primates. In E. B. Astwood (ed.), *Recent Progress in hormone research*, vol. 28, N.Y., Academic Press.
Graham-Jones, O. and Hill, W.C.O. 1962. Pregnancy and parturition in a Bornean orang. *Proc. Zool. Soc. London*, **139**, 503–510.
Grant, J. 1828. Account of the structure, manners, and habits of an orang-outang from Borneo, in the possession of George Swinton, Esq., Calcutta. *Edinb. J. Sci.* **9**, 1–24.
Green, K. F. and Garcia, J. 1971. Recuperation from illness: flavor enhancement for rats. *Science*, **173**, 749–751.

Groos, K. 1898. *The Play of Animals*. New York: D. Appleton and Co.
Groves, C. P. 1971. Pongo pygmaeus. *Mammalian Species*, **4**, 1–6.
Gustafson, C. R., Brett, L. P., Garcia, J., and Kelly, D. J. 1978. A working model and experimental solutions to the control of predatory behavior. In Markowitz, H. and Stevens, V. J. (eds.), *Behavior of Captive Wild Animals*, pp. 21–66.
Haggerty, M. E. 1913. Plumbing the minds of apes. *McClure's Mag.* **41**, 151-154.
Hahn, E. 1968. *On the Side of the Apes*. New York, Crowell.
Hamburg, D. A. and McCown, E. R., (eds.), 1979. *The Great Apes.* Menlo Park, The Benjamin/Cummings Publ. Co.
Hamilton, W. J., Buskirk, W. E. and Buskirk, W. H. 1975. Defensive stoning of baboons. *Nature*, **256**, 488–489.
Harlow, H. F. 1971. *Learning to Love.* Albion Publishing Co., San Francisco.
Harlow, H. F. 1962. The heterosexual affectional system in monkeys. *Amer. Psychologist*, **17**, 1–9.
Harlow, H. F. and Harlow, M. K. 1962. Social deprivation in monkeys. *Sci. Am.*, **207**, 137–146.
Harlow, H. F. and Harlow, M. K. 1971. Psychopathology in monkeys. In H. D. Kimmel (ed.), *Experimental Psychopathology: Recent Research and Theory*, New York, Academic Press, 203–229.
Harrisson, B. 1961. Orang-utan: what chances of survival? *Sarawak Museum Journal,* **X**, 17-18, pp. 238-261.
Harrisson, B. 1962. First response to freedom of young Orang-utans at Bako National Park, Sarawak. Unpublished report, 17 pp.
Harrisson, B. 1962. *Orang-utan.* Collins, London.
Harrisson, B. 1969. The nesting behaviour of semi-wild juvenile orang-utans. *Sarawak Mus.* **J**, XVII, 336-384.
Hedeiger, H. 1950. *Wild Animals in Captivity*. London, Butterworth.
Heinrichs, W. L. and Dillingham, L. A. 1970. Bornean orang-utans born in captivity. *Folia Primatol.*, **13**, 150–154.
Hess, J. P. 1973. Some observations on the sexual behavior of captive lowland gorillas. In *Comparative Ecology and Behaviour of Primates*, R. P. Michael and J. H. Crook, eds., New York, Academic Press, 508–581.
Hill, Clyde A. 1966. Coprophagy in Apes. *International Zoo Yearbook*, , pp. 251–257.
Hobson, Bruce. The Diagnosis of Pregnancy in the Lowland Gorilla and the Sumatran Orang Utan. *Jersey Wildlife Preservation Trust Twelfth Annual Report*, 1975, pp. 71–75.
Hodgen, G. D., Turner, C. K., Smith, E. E., and Bush, R. M. 1977. Pregnancy diagnosis in the orangutan (*Pongo pygmaeus*) using the Subhuman Primate Pregnancy Test Kit. *Laboratory Animal Science,* **27**, pp. 99-101.
Hodos, W. and Campbell, C.B.G. 1969. *Scala naturae*: Why there is no theory in comparative psychology. *Psychological Review*, **76**, 337–350.
Hoff, M. P., Nadler, R. D. and Maple, T. 1977. The development of social play in a captive group of gorillas. Paper presented at the Inaugural meeting of the American Society of Primatologists, Seattle, Washington.
Hooff, J.A.R.A.M. van. 1962. Facial expressions in higher primates. *Symp. Zool. Soc. Lond.*, **8**, 97–125.

Hooff, J.A.R.A.M. van. 1973. The Arnhem Zoo chimpanzee consortium: an attempt to create an ecologically and socially acceptable habitat. *International Zoo Yearbook*, **13**, 195–205.

Hooijer, D. A. 1949. Mammalian evolution in the Quarternary of Southern and Eastern Asia. *Evolution*, **3**, 125–128.

Hornaday, W. T. 1885. Two years in the jungle. The experiences of a hunter and naturalist in India, Ceylon, the Malay Peninsula and Borneo, London, Scribners.

Hornaday, W. T. 1922. *The Minds and Manners of Wild Animals,* New York, Scribners.

Horr, D. A. 1972. The Borneo orang-utan. *Borneo Research Bulletin*, **4**, pp. 46–50.

Horr, D. A. 1978. Orang-utan Maturation: Growing up in a Female World. In Chevalier-Skolnikoff, S. and Poirier, F. E. *Primate Bio-social Development*, pp. 289–322.

Jacobsen, C., Jacobsen, M. and Yoshioka, J. 1932. Development of an infant chimpanzee during her first year. *Comp. Psychol. Monogs.*, **9**, 1–94.

Jantschke, F. 1972. *Orang-utans in Zoologischen Garten*. Munchen, R. Piper and Co.

Jones, M. L. 1977. Second captive generation reproduction of orang utans. AAZPA So. Workshop, Little Rock.

Kingsley, S. 1977. Early mother-infant behaviour in two species of great ape. *The Dodo*, **14**, 55–65.

Klös, U. and Klös, H. G. 1966. Bemerkungen zur Künstlichen Aufzucht von Orang-Utans. *S. Ber. Ges. naturf. Freunde Berl.* (N.F.), **6**, 66–76.

Köhler, W. 1925. *The Mentality of Apes*. London, Routledge and Kegan Paul.

Kortlandt, A. and Kooij, M. 1963. Protohominid behaviour in Primates. *Symp. Zool. Soc. Lond.*, **10**, 61–88.

Kurt, F. 1970. Final report to IUCN-WWF Int. of Proj. 596, Leuser Reserve (Sumatra), 1–111.

Lethmate, J. 1977a. Versuche zum Schlagstockverfahren mit zwei jungen Orang-utans. *Zool. Anz. Jena,* **199**, 314, 209–226.

Lethmate, J. 1977b. Werkzeugherstellung eines jungen orang-utans. *Behaviour*, LX11, 1–2, 174–189.

Lethmate, J. 1977c. Nestbauverhalten eines isoliert aufgezogenen jungen Orang-utans. *Primates,* 18, 3, 545–554.

Lethmate, J. 1977d. Weitere Versuche zum Manipulier–und Werkzeugverhalten junger Orang-Utans. *Primates*, **18**, 3, 531–543.

Lethmate, J. 1978. Versuche zum "vorbedingten" Handeln mit einem jungen Orang-Utan, *Primates*, **19**, 4, 727–736.

Lethmate, J. and Dücker, G. 1973. Untersuchungen zum Selbsterkennen im Spiegel bei Orang-Utans und einigen anderen Affenarten. *Zeitschrift für Tierpsychologie*, **33**, 248–269.

Lindburg, D. G. 1973. Grooming behavior as a regulator of social interactions in rhesus monkeys. In C. R. Carpenter and D. K. Candland, eds., *Behavioral Regulators of Behavior in Primates*, Lewisburg, Bucknell University Press.

Lippert, W. 1974. Beobachtungen zum Schwangerschafts und Geburtsverhalten beim Orang utan (*P. pgymaeus*) im Tierpark Berlin. *Folia primat.* **21**, 108–134.

Loizos, C. 1967. Play behaviour in higher primates: a review. In *Primate Ethology*, D. Morris, ed., Chicago, Aldine, 176–218.

MacKinnon, J. 1971. The orang-utan in Sabah today. *Oryx*, **11**, 141–191.

MacKinnon, J. 1973. Orang-utans in Sumatra. *Oryx*, **12**, 234–242.

MacKinnon, J. 1974. The Behaviour and Ecology of wild Orang-utans (*Pongo Pygmaeus*). *Anim. Behav.*, **22**, 3–74.

MacKinnon, J. 1974. *In Search of the Red Ape*, New York, Holt, Rinehart, Winston.

MacKinnon, J. 1977. The future of orang-utans. *New Scientist*, **74**, 1037, pp. 697–699.

MacKinnon, J. 1979. Reproductive behavior in wild orangutan populations. In Hamburg, D. A. and McCown, E. A., Eds., *The Great Apes*, pp. 257–274.

Maple, T. 1974. *Basic studies of interspecies attachment behavior.* Unpublished Doctoral Dissertation, University of California at Davis.

Maple, T. 1977. Unusual Sexual Behaviors of Nonhuman Primates. In *Handbook of Sexology*, J. Money and H. Musaph, eds., Amsterdam, Elsevier.

Maple, T. 1979. Great Apes in Captivity: the Good, the bad, and the Ugly. In Erwin, J., Maple, T., and Mitchell, J., eds., *Captivity and Behavior in Primates.* New York, Van Nostrand Reinhold.

Maple, T. and Westlund, B. 1975. The integration of social interactions between cebus and spider monkeys in captivity. *Appl. Anim. Ethol.*, **1**, 305–308.

Maple, T. and Westlund, B. 1977. Interspecies dyadic attachment before and after group experience. *Primates*, **18**, 379–386.

Maple, T., Wilson, M. E., Zucker, E. L. and Wilson, S. F. 1978. Notes on the Development of a mother-reared orang-utan: the first six months. *Primates*, **19**, 3, 593–602.

Maple, T. and Zucker, E. L. 1978. Ethological studies of play behavior in captive great apes. In *Social Play in Primates*, E. O. Smith, ed., New York, Academic Press, 113–142.

Maple, T., Zucker, E. L., and Dennon, M. B. 1979. Cyclic proceptivity in a captive female orang-utan. *Behavioural Processes*, **4**, 53–59.

Markowitz, H. 1979. Environmental Enrichment and Behavioral Engineering for Captive Primates. In Erwin, J., Maple, T., and Mitchell, G. (eds.), *Captivity and Behavior: Primates in Breeding Colonies, Laboratories and Zoos.* N.Y., Van Nostrand Reinhold, 217–238.

Markowitz, H. and Woodworth, G. 1978. Experimental Analysis and Control of Group Behavior. In Markowitz, H. and Stevens, V. J. (eds.), *Behavior of Captive Wild Animals*, Chicago, Nelson-Hall, pp. 107–132.

Marler, P. 1965. Communication in Monkeys and Apes. In DeVore, I., ed., *Primate Behavior.* New York, Holt, Rinehart, Winston, 544–584.

Mason, W. A. 1963. The effects of environmental restriction on the social development of rhesus monkeys. In *Primate Social Behavior*, C. H. Southwick (ed.), Princeton, N.J., D. Van Nostrand Co., pp. 161–173.

Mason, W. A. and Green, P. C. 1972. The effects of social restriction on the behavior of rhesus monkeys: IV. responses to a novel environment and to an alien species. *J. Comp. Physio. Psych.*, **55**, 3, 363–368.

McGinnis, P. 1973. Patterns of sexual behavior in a community of free-living chimpanzees. Unpublished Doctoral Dissertation, Stanford University.

Mears, C. E. and Harlow, H. F. 1975. Play: early and eternal. *Proc. Nat. Acad. Sci.*, **72**, 1878–1882.

Medway, L. 1959. Food bone in Niah cave excavations. *Sarawak Mus. J.*, XIII, 627–636.

Menzel, E. W. 1969. Naturalistic and experimental approaches to primate behavior, in *Naturalistic Viewpoints in Psychological Research*, E. P. Willems and H. L. Rausch, eds., New York, Holt, Rinehart, and Winston, 78–121.

Menzel, E. W., Jr., Davenport, R. K., and Rogers, C. M. The Development of tool-using in wild-born and restriction reared chimpanzees. *Folia primat.*, 1970, **12**, 273–283.

Menzel, E. W. and Draper, W. A. 1965. Primate selection of food by size. *Journal of Comparative and Physiological Psychology*, **59**, 2, 231–239.

Mitchell, G. 1979. *Behavioral Sex Differences in Nonhuman Primates*. N.Y., Van Nostrand Reinhold.

Murphy, D. E. 1976. Enrichment and occupational devices for orang utans and chimpanzees. *International Zoo News*, **23.5**, 137. pp. 24–26.

Naaktgeboren, C. 1971. Die Geburt bei Haus- und Wildhunden. Neue Brehm-Bücherei, Heft 436 (Ziemsen-Verlag, Wittenberg Lutherstadt).

Nadler, R. D. 1976. Sexual behavior of captive lowland gorillas. *Arch. Sex. Behav.*, **5**, 487–502.

Nadler, R. D. and Braggio, J. T. 1974. Sex and species differences in captive-reared juvenile chimpanzees and orang-utans. *Journal of Human Evolution*, **3**, 541–550.

Nadler, R. D. 1976. Sexual behavior of the chimpanzee in relation to the gorilla and orang-utan. In Bourne, G. H. (ed.), *Progress in Ape Research*, 191–206.

Nadler, R. D. 1977. Sexual behavior of captive orangutans. *Arch Sex. Behav.*, **6**, 6, 457–475.

Nadler, R. D. and Tilford, B. 1977. Agonistic Interactions of Captive Female Orang-utans with Infants. *Folia Primat.*, **28**, 298–305.

Napier, J. R. and Napier, P. H. 1967. *A handbook of Living Primates*. London, Academic Press.

Niemitz, C. and Kok, D. 1976. Observations on the vocalization of a captive infant Orang utan (Pongo pygmaeus). *Sarawak Museum Journal*, **24**, 237–250.

Nissen, H. W. 1928. Sex development in Apes. *Comp. Psych. Monogr.*, **5**, 1, 1–165.

Oki, J. and Maeda, Y. 1973. Grooming as a regulator of behavior in Japanese macaques. In *Behavioral Regulators of Behavior in Primates*, C. R. Carpenter (ed.), New Jersey, Bucknell University Press, 149–163.

Palthe, T.V.W. and van Hooff, J.A.R.A.M. 1975. A case of adoption of an infant chimpanzee by a suckling foster chimpanzee. *Primates*, **16**, 2, 231–234.

Parker, C. 1969. Responsiveness, manipulation, and implementation behavior in chimpanzees, gorillas, and orang-utans. *Proc. 2nd Int. Congr, Primat.*, **1**, 160–166. N.Y., S. Karger.

Perry, J. 1976. Orang-utans in Captivity. *Oryx*, XIII, 3, 262–264.

Perry, J. and Horsemen, D. L. 1972. Captive Breeding of Orangutans. *Zoologica*, 105-108.

Puleo, S. G., Zucker, E. L., and Maple, T. in press. Social rehabilitation and foster mothering in capitve orang-utans. *Zool Garten.*

Rennie, J. 1838. The menageries. Library of Entertaining Knowledge, London, 442.

Reuther, R. 1966. Barrier dimensions for lions, tigers, bears, and great apes. *International Zoo Yearbook*, **16**: 217–222.

Rijksen, H. D. 1978. A Field Study on Sumatran Orang Utans (*Pongo pygmaeus abelii* Lesson 1827). Wageningen, H. Veenman and Zonen B. V., 420 pps.

Rijksen, H. D. and Rijksen-Graatsma, A. G. 1975. Orang Utan Rescue Work in North Sumatra. *Oryx*, **XIII**, 1, 63–73.

Rock, M. A. 1977. Orang–Endangered "Man of the Forest." *National Parks & Conservation Magazine*, **51**, 8, 10–15.

Rodman, P. S. 1973. Population composition and adaptive organization among orang-utans of the Kutai Reserve. In *Comparative Ecology and Behaviour of Primates*, J. H. Crook and R. P. Michael, eds., London, Academic Press, 171–209.

Rodman, P. S. 1977. Feeding behaviour of orang-utans of the Kutai Nature Reserve, East Kalimantan. In *Primate Ecology*, T. H. Clutton-Brock, ed., New York, Academic Press, 383–413.

Rodman, P. S. 1979. Individual activity patterns and the solitary nature of orangutans. In Hamburg, D. A. and McCown, E. R., Eds., *The Great Apes,* 235-255.

Rogers, C. M. and Davenport, R. K. 1969. Effects of restricted rearing on sexual behavior of chimpanzees. *Develop. Psychol.,* **1**, 200–204.

Rosenblum, L. A., Kaufman, I. C. and Stynes, A. J. 1966. Some characteristics of adult social and autogrooming patterns in two species of macaque. *Folia Primat.* **4**, 438–451.

Sade, D. S. 1965. Some aspects of parent-offspring and sibling relations in a group of rhesus monkeys, with a discussion of grooming. *Amer. J. Phys. Anthrop.*, **23**, 1–17.

Rumbaugh, D. M. 1970. Learning skills of anthropoids. In *Primate Behavior*, L. A. Rosenblum, ed., New York, Academic Press, 1–70.

Rumbaugh, D. M. and Gill, T. V. 1970. The learning skills of Pongo. *Proc. 3rd Congr. Int. Prim. Soc.*, Basel, Karger, 158–163.

Rumbaugh, D. M., Gill, T. V., and Wright, S. C. 1973. Readiness to attend to visual foreground cues. *Journal of Human Evolution*, **2**, 181–188.

Rumbaugh, D. M. and McCormack, 1967. The learning skills of primates: a comparative study of apes and monkeys. In D. Stark, R. Schneider, and H. J. Kuhn (eds.). *Progress in Primatology*, Stuttgart: Gustav Fischer.

Rumbaugh, D. M., Riesen, A. H., and Wright, S. C. 1972. Creative responsiveness to objects: a report of a pilot study with young apes. *Folia primatologica*, **17**, 397–403.

Savage, E. S. and Malick, C. 1976. Play and socio-sexual behavior in a captive chimpanzee group. *Behaviour*, **60**, 179–194.

Schaller, G. 1961. The orang-utan in Sarawak. *Zoologica*, **46**, 73–82.

Schiller, P. 1957. Innate motor actions as a basis of learning. In C. H. Schiller (ed.), *Instinctive Behavior*, New York, Int. Univ. Press, 264–287.

Schmidt, M. 1878. Beobachtungen am Orang-utan. *Zool. Gart.*, Frankfurt, 19: 81–82.

Schrier, A. M., Harlow, H. F. and Stollnitz, F. (eds.). 1965. *Behavior of Nonhuman Primates*. New York, Academic Press.

Schultz, A. H. 1941. Growth and development of the orang-utan. *Contributions to Embryology*, Carnegie Institute Washington Publication Number 525, **29**, 57–110.

Schultz, A. H. 1938. Genital swelling in the female orang-utan. *J. Mammal.*, **19**, 363–366.

Schultz, A. H. 1969. *The Life of Primates*. London, Academic Press.

Scollay, P. A., Joines, S., Baldridge, C., and Cuzzone, A. 1975. Learning to be a mother. *Zoonooz*, **XLVIII**, 4, 4–9.

Seitz, A. 1969. Einige Feststellungen zur Pflege und Aufzucht von Orang-utans, *Pongo pygmaeus* Hoppius 1763. *Zool. Garten, 36*, 225–245.

Senko, M. G. 1966. The effects of early, intermediate, and late experience upon adult macaque sexual behavior. M. S. Thesis, University of Wisconsin, Madison.

Sheak, W. H. 1922. Disposition and intelligence of the orang-utan. *J. Mammal.*, **3**, 47–51.

Simons, E. L. 1971. Hipparion in Tertiary siwaliks. *Nature* (*Lond.*), 229, 408–409.

Smith, E. O., ed. 1978. *Social Play in Primates*. New York: Academic Press.

Snyder, R. L. 1976. Strategies for feeding captive omnivorous animals. *Proc. AAZPA Conf.*, 127–140.

Sommer, R. 1974. *Tight Spaces*. Englewood Cliffs, N.J., Prentice-Hall.

Sonntag, C. F. 1924. *The morphology and evolution of the apes and man*. London, 364 pps.

Suomi, S. J. 1973. Surrogate rehabilitation of monkeys reared in total isolation. *J. Child Psychol. Psychiat.*, **14**, 71–77.

Suomi, S. and Harlow, H. F. 1972. Social rehabilitation of isolate-reared monkeys. *Develop. Psych.*, **6**, 3, 487–496.

Taub, D. M., Lehrner, N.D.M. and Adams, M. R. 1977. Enforced adoption and successful raising of a neonate squirrel monkey Saimiri sciureus. *Lab. Prim. Newsl.*, **16**, 810.

Thorington, R. W. 1978. Conservation and Primatology: A glum view of the future. In Chivers, D. J. and Lane-Petter, W., Eds. *Recent Advances in Primatology*, London, Academic Press, pp. 3–6.

Tigges, J. 1963. On color vision in Gibbon and Orang-utan. *Folia Primat.*, **1**, 188–198.

Turner, C. H., Davenport, R. K., and Rogers, C. M. 1969. The effect of early deprivation on the social behavior of adolescent chimpanzees. *Amer. J. Psychiat.*, **125**, 11, 85–90.

Tuttle, R. H. and Rogers, C. M. 1966. Genetic and selective factors in reduction of the hallux in *Pongo pygmaeus*. *Am. J. Phys. Anthro.*, **24**, 191–198.

Ullrich, W. 1970. Geburt and naturliche Geburtshilfe beim Orang-utan. *Zool. Garten*, 284–289.

Van Lawick-Goodall, J. Tool-using in primates and other vertebrates. In Lehrman, D. S. et al., eds., *Advances in the Study of Behavior*, 1970, N.Y., Academic Press, **3**, 195–249.

Wallace, A. R. 1856. Some Account of an infant orang-utang. *Annals and Magazine of Natural History*, **17**, pp. 386–390.
Wallace, A. R. 1869. *The Malay Archipelago*. London, Macmillan.
Washburn, S. and DeVore, I. 1961. The Social life of baboons. *Sci. Amer.*, **204**, 62–71.
White, L. 1978. Behavioral response of orangutans to remodeling of their exhibit. *Animal Keeper's Forum*, **5**, 11, 179–180.
Wilson, M. L. and Elicker, J. G. 1976. Establishment, maintenance, and behavior of free-ranging chimpanzees on Ossabaw Island, Georgia U.S.A. *Primates*, **17**, 451–473.
Wilson, M. E., Maple, T., Nadler, R. D., Hoff, M. P., and Zucker, E. L. 1977. Characteristics of paternal behavior in captive orang-utans and lowland gorillas. Paper presented at the Inaugural meeting of the American Society of Primatologists, Seattle, Washington.
Wrangham, R. W. 1974. Artificial feeding of chimpanzees and baboons in their natural habitat. *Anim. Behav.*, **22**, 83–93.
Yerkes, R. M. 1925. *Almost Human*. New York, The Century Co.
Yerkes, R. M. 1927. The Mind of a Gorilla, Part 1, *Genetic Psych. Monogs.*, **2**, 1–2, 193 pp.
Yerkes, R. M. 1933. Genetic aspects of grooming, a socially important primate behavior pattern. *J. Social Psychol.*, **4**, 3–25.
Yerkes, R. M. 1943. *Chimpanzees: A Laboratory Colony*. New Haven, Yale University Press.
Yerkes, R. M. and Yerkes, A. W. 1929. *The Great Apes*. New Haven, Yale University Press.
Zucker, E. L., Brogdor, L., and Maple, T. 1976. Social behavior of captive orang-utans. Paper presented at conference of Southeastern Psychological Association, New Orleans, La.
Zucker, E. L., Puleo, S. and Maple, T. 1977. The development of sexual behaviors through play in captive young orang-utans. Paper presented at SEPA meeting, Hollywood, Fla.
Zucker, E. L., Stine, W. W., Hoff, M. P., Nadler, R. D., Dennon, M. B., and Maple, T. In prep. Grooming behaviors of orang-utans and gorillas: description and comparison.
Zuckerman, S. 1933. *Functional Affinities of Man, Monkeys, and Apes*. New York, Harcourt, Brace and Company

Additional References

Borner, M. and Gittens, P. 1978. Round table discussion on rehabilitation. In: D. J. Chivers and W. Lane-Petter (eds.), *Recent Advances in Primatology*. London, Academic Press, pp. 101-106.
Breland, K. and Breland, M. 1961. The misbehavior of organisms. *Amer. Psychol.*, **16**, 681-684.

Ellenberger, H. F. 1964. The mental hospital and the zoological garden. In: J. and B. Klaits (eds.), *Animals and Men in Historical Perspective.* New York, Harper and Row, pp. 59–92.

Grand, T. I. 1972. A mechanical interpretation of terminal branch feeding. *J. Mammal.,* **53**, 198–201.

Kummer, H. 1971. *Primate Societies.* Chicago, Aldine.

Lomax, E. M. 1978. *Science and Patterns of Child Care.* San Francisco, W. H. Freeman Co.

Martin, R. D., Kingsley, S. R. and Stavy, M. 1977. Prospects for coordinated research into breeding of great apes in zoological collections. *The Dodo,* **14**, 45–55.

Mitchell, G. 1968. Persistent behavior pathology in rhesus monkeys following early social isolation. *Folia Primat.,* **8**, 132–147.

Thorington, R. W. 1978. Conservation and primatology: a glum view of the future. In: D. J. Chivers and W. Lane-Petter (eds.), *Recent Advances in Primatology.* London, Academic Press, pp. 3–6.

Wilson, E. O. 1975. *Sociobiology: The New Synthesis.* Cambridge, Massachusetts, Harvard University Press.

Name Index

Subject Index